T0325228

Design for Reliability

WILEY SERIES IN QUALITY & RELIABILITY ENGINEERING
*and related titles**

Electronic Component Reliability:
Fundamentals, Modelling, Evaluation and Assurance
Finn Jensen

Measurement and Calibration Requirements
For Quality Assurance to ASO 9000
Alan S. Morris

Integrated Circuit Failure Analysis:
A Guide to Preparation Techniques
Friedrich Beck

Test Engineering
Patrick D. T. O'Connor

Six Sigma: Advanced Tools for Black Belts and Master Black Belts*
Loon Ching Tang, Thong Ngee Goh, Hong See Yam, Timothy Yoap

Secure Computer and Network Systems: Modeling, Analysis and Design*
Nong Ye

Failure Analysis:
A Practical Guide for Manufacturers of Electronic Components and Systems
Marius Bâzu and Titu Băjenescu

Reliability Technology:
Principles and Practice of Failure Prevention in Electronic Systems
Norman Pascoe

Effective FMEAs: Achieving Safe, Reliable, and Economical Products and Processes
Using Failure Mode and Effects Analysis
Carl Carlson

Design for Reliability
Dev Raheja and Louis J. Gullo (Editors)

Design for Reliability

Edited by

Dev Raheja

Louis J. Gullo

A JOHN WILEY & SONS, INC., PUBLICATION

Published by John Wiley & Sons, Inc., Hoboken, New Jersey.
Published simultaneously in Canada.

For general information on our other products and services or for technical support, please contact our Customer Care Department within the United States at (800) 762-2974, outside the United States at (317) 572-3993 or fax (317) 572-4002.

Wiley also publishes its books in a variety of electronic formats. Some content that appears in print may not be available in electronic formats. For more information about Wiley products, visit our web site at www.wiley.com.

Library of Congress Cataloging-in-Publication Data:

Raheja, Dev.
Design for reliability / Dev Raheja & Louis J. Gullo.
p. cm.
ISBN 978-0-470-48675-7 (hardback)
1. Reliability (Engineering) I. Gullo, Louis J. II. Title.
TA169.R348 2011
620′.00452–dc23

2011042405

Printed in the United States of America

10 9 8 7 6 5 4 3 2 1

To my wife, Hema, and my children, Gauri, Pramod, and Preeti
Dev Raheja

To my wife, Diane, and my children, Louis, Jr., Stephanie, Catherine, Christina, and Nicholas
Louis J. Gullo

Contents

Contributors

Steven S. Austin
Missile Defense Agency
Department of Defense
Huntsville, Alabama

Lawrence Bernstein
Stevens Institute of Technology
Hoboken, New Jersey

Joseph A. Childs
Missiles and Fire Control
Lockheed Martin Corporation
Orlando, Florida

Jack Dixon
Dynamics Research Corporation
Orlando, Florida

Louis J. Gullo
Missile Systems
Raytheon Company
Tucson, Arizona

Samuel Keene
Keene and Associates, Inc.
Lyons, Colorado

Brian Moriarty
Engility Corporation
Lake Ridge, Virginia

Dev Raheja
Raheja Consulting, Inc.
Laurel, Maryland

Robert W. Stoddard
Six Sigma IDS, LLC
Venetia, Pennsylvania

C.M. Yuhas

Foreword

The importance of quality and reliability to a system cannot be disputed. Product failures in the field inevitably lead to losses in the form of repair cost, warranty claims, customer dissatisfaction, product recalls, loss of sales, and in extreme cases, loss of life. Thus, quality and reliability play a critical role in modern science and engineering and so enjoy various opportunities and face a number of challenges.

As quality and reliability science evolves, it reflects the trends and transformations of technological support. A device utilizing a new technology, whether it be a solar power panel, a stealth aircraft, or a state-of-the-art medical device, needs to function properly and without failure throughout its mission life. New technologies bring about new failure mechanisms (chemical, electrical, physical, mechanical, structural, etc.), new failure sites, and new failure modes. Therefore, continuous advancement of the physics of failure, combined with a multi-disciplinary approach, is essential to our ability to address those challenges in the future.

In addition to the transformations associated with changes in technology, the field of quality and reliability engineering has been going through its own evolution: developing new techniques and methodologies aimed at process improvement and reduction of the number of design- and manufacturing-related failures.

The concept of design for reliability (DFR) has been gaining popularity in recent years and its development is expected to continue for years to come. DFR methods shift the focus from reliability demonstration and the outdated "test-analyze-fix" philosophy to designing reliability into products and processes using the best available science-based methods. These concepts intertwine with probabilistic design and design for six sigma (DFSS) methods, focusing on reducing variability at the design and manufacturing levels. As such, the industry is expected to increase the use of simulation techniques, enhance the applications of reliability modeling, and integrate reliability engineering earlier and earlier in the design process. DFR also transforms the role of the reliability engineer from being focused primarily on product test and analysis to being a mentor to the design team, which is responsible for finding and applying the best design methods to achieve reliability. A properly applied DFR process ensures that pursuit of reliability is an enterprise-wide activity.

Several other emerging and continuing trends in quality and reliability engineering are also worth mentioning here. For an increasing number of applications, risk assessment will enhance reliability analysis, addressing not only the probability of failure but also the quantitative consequences of that failure. Life-cycle engineering concepts are expected to find wider applications in reducing life-cycle risks and minimizing the combined cost of design, manufacturing, quality, warranty, and service. Advances in prognostics and health management will bring about the development of new models and algorithms that can predict the future reliability of a product by assessing the extent of degradation from its expected operating conditions. Other advancing areas include human and software reliability analysis.

Additionally, continuous globalization and outsourcing affect most industries and complicate the work of quality and reliability professionals. Having various engineering functions distributed around the globe adds a layer of complexity to design coordination and logistics. Moving design and production into regions with little knowledge depth regarding design and manufacturing processes, with a less robust quality system in place and where low cost is often the primary driver of product development, affects a company's ability to produce reliable and defect-free parts.

Despite its obvious importance, quality and reliability education is paradoxically lacking in today's engineering curriculum. Few engineering schools offer degree programs or even a sufficient variety of courses in quality or reliability methods. Therefore, a majority of quality and reliability practitioners receive their professional training from colleagues, professional seminars, and from a variety of publications and technical books. The lack of formal education opportunities in this field greatly emphasizes the importance of technical publications for professional development.

The real objective of the Wiley Series in Quality & Reliability Engineering is to provide a solid educational foundation for both practitioners and researchers in quality and reliability and to expand the reader's knowledge base to include the latest developments in this field. This series continues Wiley's tradition of excellence in technical publishing and provides a lasting and positive contribution to the teaching and practice of engineering.

ANDRE KLEYNER

Editor
Wiley Series in Quality & Reliability Engineering

Preface

Design for reliability (DFR) has become a worldwide goal, regardless of the industry and market. The best organizations around the world have become increasingly intent on harvesting the value proposition for competing globally while significantly lowering life cycle costs. The DFR principles and methods are aimed proactively to prevent faults, failures, and product malfunctions, which result in cheaper, faster, and better products. In Japan, this tool is used to gain customer loyalty and customer trust. However, we still face some challenges. Very few engineering managers and design engineers understand the value added by design for reliability; they often fail to see savings in warranty costs, increased customer satisfaction, and gain in market share.

These facts, combined with the current worldwide economic challenges, have created perfect conditions for this science of engineering. This is an art also because many decisions have to be made not only on evidence-based data, but also on engineering creativity to design out failure at lower costs. Readers will be delighted with the wealth of knowledge because all contributors to this book have at least 20 years hands-on experience with these methods.

The idea for this book was conceived during our participation in the IEEE Design for Reliability Technical Committee. We saw the need for a DFR volume not only for hardware engineers, but also for software and system engineers. The traditional books on reliability engineering are written for reliability engineers who rely more on statistical analysis than on improvements in inherent design to mitigate hardware and software failures. Our book attempts to fill a gap in the published body of knowledge by communicating the tremendous advantages of designing for reliability during very early development phase of a new product or system. This volume fulfills the needs of entry-level design engineers, experienced design engineers, engineering managers, as well as the reliability engineers/managers who are looking for hands-on knowledge on how to work collaboratively on design engineering teams.

ACKNOWLEDGMENTS

We would like to thank the IEEE Reliability Society for sowing the seed for this book, especially the encouragement from a former society president,

Dr. Samuel Keene, who also contributed chapters in the book. We would like to recognize a few of the authors for conducting peer reviews of several chapters: Joe Childs, Jack Dixon, Larry Bernstein, and Sam Keene. We also thank the guest editors—Tim Adams, at NASA Kennedy Center, and Dr. Nat Jambulingam, at NASA Goddard Space Flight Center—who helped edit several chapters. We are grateful to Diana Gialo, at Wiley, who has always been gracious in helping and guiding us.

We acknowledge the contributions of the following:

Steve Austin (Chapter 12)

Larry Bernstein (Chapter 13)

Joe Childs (Chapters 2, 6, and 15)

Jim Dixon (Chapters 9 and 16)

Lou Gullo (Chapters 4, 5, 10, 11, 14, and 18)

Sam Keene (Chapters 3 and 8)

Brian Moriarty (Chapter 17)

Dev Raheja (Chapter 1)

Bob Stoddard (Chapter 7)

C. M. Yuhas (Chapter 13)

Dev Raheja
Louis J. Gullo

Introduction: What You Will Learn

Chapter 1 Design for Reliability Paradigms (Raheja)

This chapter introduces what it means to design for reliability. It shows the technical gaps between the current state-of-art and what it takes to design reliability as a value proposition for new products. It gives real examples of how to get high return on investment to understand the art of design for reliability. The chapter introduces readers to the deeper level topics with eight practical paradigms for best practices.

Chapter 2 Reliability Design Tools (Childs)

This chapter summarizes reliability tools that exist throughout the product's life cycle from creation, requirements, development, design, production, testing, use, and end of life. The need for tools in understanding and communicating reliability performance is also explained. Many of these tools are explained in further detail in the chapters that follow.

Chapter 3 Developing Reliable Software (Keene)

This chapter describes good design practices for developing reliable software embedded in most of the high technology products. It shows how to prevent software faults and failures often inherent in the design by applying evidence-based reliability tools to software such as FMEA, capability maturity modeling, and software reliability modeling. It introduces the most popular software reliability estimation tool CASRE (*C*omputer *A*ided *S*oftware *R*eliability *E*stimation).

Chapter 4 Reliability Models (Gullo)

This chapter is on reliability modeling, one of the most important tools for design for reliability in he early stages of design, to determine strategy for overall reliability. The chapter covers models for system reliability, component reliability, and shows the use of block diagrams in modeling. It discusses reliability growth process, similarity analysis used for physical modeling, and widely used models for simulation.

Chapter 5 Design Failure Modes, Effects, and Criticality Analysis (Gullo)

This chapter on FMECA contains the core knowledge for reliability analysis at system level, subsystem level, and component level. The chapter shows how to perform risk assessment using a risk index called risk priority number and shows how to eliminate single-point failures, making a design significantly less vulnerable. It explains the difference between FMEA and FMECA and how to us them for improving product performance and the maintenance effectiveness.

Chapter 6 Process Failure Modes, Effects, and Criticality Analysis (Childs)

The preceding chapter showed how to make design more robust. This chapter applies the FMEA tool to analyze a process for robustness, such that the manufacturing defects are eliminated before the show up in production. The end result is improved product reliability with lower manufacturing costs. It covers step-by-step procedure to perform the analysis, including the risk assessment using the risk priority number.

Chapter 7 FMECA Applied to Software Development (Stoddard)

The FMEA tool is just as applicable to software design. There is very little literature on how to apply it to software. This chapter shows the details of how to use it to improve the software reliability. It covers the lessons learned and shows different ways of integrating the FMECA into the most widely used software development model known as "V" model. The chapter describes roles and responsibilities for proper use of this tool.

Chapter 8 Six Sigma Approach to Requirements Development (Keene)

In this chapter the author explains why design of experiments (DOE) is a sweet spot for identifying the key input variables to a six sigma programs. The chapter covers the origin of this program, the meaning of six sigma measurements, and how it is applied to improve the design. It then proceeds to cover the tools for designing the product for six sigma performance to reduce failure rates as close to zero as possible.

Chapter 9 Human Factors in Reliable Design (Dixon)

Humans are often blamed for many product failures when in fact the fault lies in the insufficient attention to human factor engineering. This chapter covers the principles of human-centered design to make man–machine interface robust and error-tolerant. It covers how to perform the human factors analysis, and how to integrate it to make the product design user-friendly.

Chapter 10 Stress Analysis During Design to Eliminate Failures (Gullo)

This chapter explains why it is critical to reduce the design stress to improve durability, as well as reliability. It introduces the concept of derating as a design tool. The author includes examples on electrical and mechanical stress analysis, including how to apply this theory to software design. The chapter also shows how to apply finite element analysis, a numerical technique, to solve specific design problems.

Chapter 11 Highly Accelerated Life Testing (Gullo)

Usually designers cannot predict what failures will occur for a new design. This chapter shows how highly accelerated life tests and highly accelerated stress tests can reveal the failure modes quickly. It covers how to design these tests and how to estimate the design margin from the test results. It shows different methods of accelerating the stresses.

Chapter 12 Design for Extreme Environments (Austin)

When a product is used in extreme cold or extreme heat, such as in Alaska or in a desert in Arizona, we must design for such environments to assure product can last long enough. This chapter shows what factors need to be considered and how to design for each condition. It shows how lessons learned from space programs and overseas experience can help make products durable, reliable, and safe.

Chapter 13 Design for Trustworthiness (Bernstein and Yuhas)

This is a very important chapter because software design methods for reliability are not standardized yet. This chapter goes beyond reliability to design software, such that it is also safe and secure from errors in engineering changes which are very frequent. This chapter covers design methods and offers suggestions for improving the architecture, modules, interfaces, and using right policies for re-using the software. The chapter offers good design practices.

Chapter 14 Prognostics and Health Management Capabilities to Improve Reliability (Gullo)

Design for reliability practices should include detecting a malfunction before a product malfunctions. This chapter covers designing prognostics and product health monitoring principles that can be designed into the product. The result is enhanced system reliability. The chapter includes condition-based maintenance and time-based maintenance, use of failure precursors to signal an imminent failure event, and automatic stress monitoring to enhance prognosis.

Chapter 15 Reliability Management (Childs)

This chapter provides both motivation and guidance in outlining the importance of good reliability management. Management participation is the key to any successful reliability in design. It shows how to manage, plan, execute, and document the needs of the program during early design. It describes the important tasks, and closing the feedback loops after reliability assessment, problem solving, and reliability growth testing.

Chapter 16 Risk Management, Exception Handling, and Change Management (Dixon)

Many risks are overlooked in a product design. This chapter defines what is risk in engineering terms, how to predict risk, assess risk, and mitigate it. It highlights the role of risk management culture in mitigating risks and the critical role of configuration management for avoiding new risks from design changes. Included in this chapter is how to minimize oversights and omissions, including requirement creeps.

Chapter 17 Integrating Design for Reliability with Design for Safety (Moriarty)

This chapter integrates reliability with safety, including how to design for safety. It covers several safety analysis techniques that equally apply to reliability. It shows the how a risk assessment code matrix is used widely in aerospace and many commercial products to make risk management decisions. It includes examples of risk reduction.

Chapter 18 Organizational Reliability Capability Assessment (Gullo)

This chapter describes the benefits of using IEEE 1624–2008 standard to describe how reliability capability of an organizational entity is determined by assessing eight key reliability practices and associated metrics. Management should know the capability of an organization to deliver a reliable product, which is defined as organizational reliability capability. It describes the process in detail with case studies.

Chapter 1

Design for Reliability Paradigms

Dev Raheja

WHY DESIGN FOR RELIABILITY?

The science of reliability has not kept pace with user expectations. Many corporations still use MTBF (mean time between failures) as a measure of reliability, which, depending on the statistical distribution of failure data, implies acceptance of roughly 50 to 70% failures during the time indicated by the MTBF. No user today can tolerate such a high number of failures. Ideally, a user does not want any failures for the entire expected life! The life expected is determined by the life inferred by users, such as 100,000 miles or 10 years for an automobile, at least 10 years for kitchen appliances, and at least 20 years for a commercial airliner. Most commercial companies, such as automotive and medical device manufacturers, have stopped using the MTBF measure and aim at 1 to 10% failures during a self-defined time. This is still not in line with users' dreams. The real question is: Why not design for zero failures if we can increase profits and gain more market share? Zero failures implies zero mission-critical failures or zero safety-critical system failures. As a minimum, systems in which failures can lead to catastrophic consequences must be designed for zero failures. There are companies that are able to do this. Toyota, Apple, Gillette, Honda, Boeing, Johnson & Johnson, Corning, and Hewlett-Packard are a few examples.

The aim of design for reliability (DFR) is to design-out failures of critical system functions in a system. The number of such failures should be

Design for Reliability, First Edition. Edited by Dev Raheja, Louis J. Gullo.

zero for the expected life of the product. Some components may be allowed to fail, such as in redundant systems. For example, in aerospace, as long as a system can function at least for the duration of the mission and the failed components are replaced prior to the next mission to maintain redundancy, certain failures can be tolerated. This is, however, insufficient for complex systems where thousands of software interactions, hundreds of wiring connections, and hundreds of human factors affect the systems' reliability. Then there are issues of compatibility [1] among components and materials, among subsystems, and among hardware and software interactions. Therefore, for complex systems we may find it impossible to have zero failures, but we must at least prevent the potential failures we know about. Since failures can come from unknown and unexpected interactions, we should try to design-in fallback modes for unexpected events. A "what-if" analysis usually points to some events of this type. To minimize failures in complex systems, in this book we describe techniques for improving software and interface reliability.

As indicated earlier, some companies have built a strong and long-lasting reputation for reliability based on aiming at zero failures. Toyota and Sony built their world leadership mostly on high reliability; and Hyundai has been offering a 10-year warranty and increasing its market share steadily. Progress has been made since then. In 1974, when nobody in the world gave a warranty longer than one year, Cooper Industries gave a 15-year warranty to electric power utilities on high-voltage transformer components and stood out as the leader in profitability among all Fortune 500 electrical companies. Raytheon has established a culture at the highest level in the corporation of providing customers with mission assurance through a "no doubt" mindset. Says Bill Swanson, chairman and CEO of Raytheon: "[T]here must be no doubt that our products will work in the field when they are needed" (Raytheon Company, *Technology Today*, 2005, Issue 4). Similarly, with its new lifetime power train warranty, Chrysler is creating new standards for reliability.

REFLECTIONS ON THE CURRENT STATE OF THE ART

Reliability is defined as the probability of performing all the functions (including safety functions) satisfactorily for a specified time and specified use conditions. The functions and use conditions come from the specification. If a specification misses or is vague 60% or more of the time, the reliability predictions are of very little value. This is usually the case [2]. The second big issue is: How many failures should be tolerable? Some readers may not agree that we can design for zero critical failures, but the evidence supports the contrary conclusion. We may not be able to prevent failures that we did not foresee, but we can design out all the critical failure modes that we discover during the requirements analysis and in the failure mode and effects analysis (FMEA). In over 30 years' experience, I have yet to encounter a failure mode that cannot be designed-out. The cost is usually not an issue if the FMEA is conducted and the improvements are made during the early design stage. The time specified for critical failures in the reliability definition should be the entire lifetime expected.

In this chapter we address how to write a good system specification and how to design so as not to fail. We make it clear that the design for reliability should concentrate on the critical and major failures. This prevents us from solving easy problems and ignoring the complex ones. The following incident raises issues that are central to designing for reliability.

> *The lessons learned from the Interstate 35 bridge collapse in Minnesota on August 1, 2007 into the Mississippi River on August 1, killing 13, give us some clues about what needs to be done. Similar failure mechanisms can be found in many large electrical and mechanical systems, such as aircraft and electric power plants.*
>
> *The bridge was expanded from four lanes to six, and eventually to eight. Some wonder whether that might have played a role in its collapse. Investigators said the failure resulted because of a flaw in its design. The designers had specified a metal plate that was too thin to serve as a junction of several girders.*
>
> *Like many products, it gradually got exposed to higher loads, adding strain to the weak spot. At the time of the collapse, the maintenance crews had brought tons of equipment and material onto the deck for a repair job. The bridge was of a design known as a nonredundant structure, meaning that if a single part failed, the entire structure could collapse. Experts say that the pigeon dung all over the steel could have caused faster corrosion than was predicted.*

This case history challenges the fundamentals of engineering taught in the universities.

- *Should the design margin be 100% or 800%?* "How does the designer determine the design margin?"
- *Should we design for pigeons doing their dirty job?* What about designing for all the other environmental stressors, such as chemicals sprayed during snow emergencies, tornados, and earthquakes?
- *Should we design-in redundancy on large mechanical systems to avoid disasters?* The wisdom says that redundancy delays failures but may not avoid disasters. The failure could occur in both the redundant paths, such as in an aircraft accident where the flying debris cut through all three redundant hydraulic lines.
- *Should we design for sudden shocks experienced by the bridge during repair and maintenance?*

These concerns apply to any product, such as electronics, electrical power systems, and even a complex software design. In software, the corrosion can be symbolic for applying too many patches without knowing the interactions. Call it "software corrosion."

The answers to the questions above should be a resounding "yes." An engineering team should foresee all these and many more failure scenarios before starting to design. The obvious strategy is to write a good system specification by first predicting all major potential failures and avoiding them by writing robust requirements. Oversights and omissions in specifications are the biggest weakness in the design for reliability. Typically, 200 to 300 requirements are generally missing or vague for a reasonably complex system such as an automotive transmission.

Analyses techniques covered in this book for hardware and software help us discover many missing requirements, and a good brainstorming session for overlooked requirements always results in discovering many more. What we really need is perhaps the paradigms based on lessons learned.

THE PARADIGMS FOR DESIGN FOR RELIABILITY

Reliability is a process. If the right process is followed, results are likely to be right. The opposite is also true in the absence of the right process. There is a saying: "If we don't know where we are going, that's where we will go." It is difficult enough to do the right things, but it is even more difficult to know what the right things are!

Knowledge of the right things comes from practicing the use of lessons learned. Just having all the facts at your fingertips does not work. One must utilize the accumulated knowledge for arriving at correct decisions. Theory is not enough. One must keep becoming better by practicing. Take the example of swimming. One cannot learn to swim from books alone; one must practice swimming. It is okay to fail as long as mistakes are the stepping stones to failure prevention. Thomas Edison was reminded that he failed 2000 times before the success of the light bulb. His answer, "I never failed. There were 2000 steps in this process."

One of the best techniques is to use lessons learned in the form of paradigms. They are easy to remember and they make good topics for brainstorming during design reviews.

Paradigm 1: Learn To Be Lean Instead of Mean

When engineers say that a component's life is five years, they usually imply the calculation of the mean value, which says that there is a 50% chance of failure during the five years. In other words, either the supplier or the customer has to pay for 50% failures during the product cycle. This is expensive for both: a lose–lose situation. Besides, there are many indirect expenses: for warranties, production testing, and more inventories to replace failed parts. This is mean management. It has a negative return on investment. It is mean to the supplier because of loss of future business and mean to the customer in putting up with the frustrations of downtime and the cost of business interruptions. Therefore, our failure rate goal should be *as lean as possible*. Engineers should promise *minimum life* to customers, not mean life. Never use averages in reliability; they are of no use to anyone.

Paradigm 2: Spend a Lot of Time on Requirement Analysis

It is worth repeating that the sources of most failures are incomplete, ambiguous, and poorly defined requirements. That is why we introduce unnecessary design changes and write deviations when we are in hurry to ship a product. Look particularly for missing functions in the specifications. There is often practically nothing in a specification about modularity, reliability, safety, serviceability, logistics, human factors, reduction of "no faults found," diagnostics capability, and prevention of warranty failures. Very few specifications address even obvious requirements, such as internal interface, external interface, user–hardware interface, user–software interface, and how the product should behave if and when a sneak failure occurs. Developing a good specification is an iterative process with inputs from the customer and the entities that are downstream in the process. Those who are trying to build reliability around a faulty specification should only expect a faulty product. Unfortunately, most companies think of reliability when the design is already approved. At this stage there is no budget and no time for major design changes. The only thing a company can do is to hope for reasonable reliability and commit to do better the next time.

To identify missing functions, a cross-functional team is necessary. At least one member from each disciple should be present, such as manufacturing, field service, and marketing, as well as a customer representative. If the specification contains only 50% of the necessary features, how can one even think of reliability? Reliability is not possible without accurate and comprehensive specifications. Therefore, writing accurate performance specifications is a prerequisite for reliability. Such specifications should aim at zero failures for the modes that result in product recalls, high downtime, and inability to diagnose. My interviews with those attending my reliability courses reveal that the dealers are unable to diagnose about 65% of the problems (no faults found). Obviously, fault isolation requirements in the specifications are necessary to reduce down time.

To ensure the accuracy and completeness of a specification, only those who have knowledge of what makes a good specification should approve it. They must ensure that the specification is clear on what the product should never do, however stupid it may sound. For example: "There shall be no sudden acceleration during landing" for an aircraft. In addition, the marketing and sales experts should participate in writing the specification to make sure that old warranty problems "shall not" be in the new product and that there is enough gain in reliability to give the product a competitive edge.

The "shall not" specification is not limited to failures. That would be too simple. We must be able to see the complexity in this simplicity. This is called *interconnectedness*. We need to know that reliability is intertwined with many elements of life-cycle costs. The costs of downtime, repairs, preventive maintenance, amount of logistics support required, safety, diagnostics, and serviceability are dependent on the level of reliability. In the same spirit, we should also analyze product friendliness and modularity, which are interconnected with reliability. For example, General Motors is designing its hydrogen cars to have a single chassis for all models instead of 80 different chassis as is the case with current production. This action influences reliability in many ways. Similarly, an analysis of downtime should be conducted by service engineering staff to ensure that each fault will be diagnosed in a timely manner, repairs will be quick, and life-cycle costs will be reduced by extending the maintenance cycles or eliminating the need for maintenance altogether. The specification should be critiqued for quick serviceability and ease of access. Until the specification is written thoroughly and approved, no design work should begin. An example of the need to identify missing requirements is that nearly 1000 people around the world lost their lives while the kinks were being removed from the 290-ton McDonnell Douglas DC-10 during the 1970s. Blown-out cargo doors, shredded hydraulic lines, and engines dropped during the flight were just a few of the behemoth's early problems. It is obvious that the company did not have the right system performance specification. We rely on customers to tell us what they want, but they themselves don't know many requirements until there is a breakdown. Customers are not going to tell us that the cargo doors should not blow out during a crowded flight. It is the design team's responsibility to figure out what the customers did not say.

To find the design flaws early, a team has to view the system from various angles. You would not buy a house by just looking at the front view. You want to see it from all sides. Similarly, a product concept has to be viewed from at least the following perspectives:

- Functions of the product
- Range of applications
- Range of environments
- Active safety
- Duty cycles during life
- Reliability
- Robustness for user or servicing mistakes
- Logistics requirements
- Manufacturability requirements
- Internal interface requirements
- External interface requirements
- Installation requirements
- Shipping and handling capabilities
- Serviceability and diagnostics capabilities
- Prognostics health monitoring
- Usability on other products
- Sustainability

There is a need to explain a sustainable design in the list above. Good product design is about meeting current needs without compromising the needs of future generations, such as by pollution or global warming. Current electronic and computers are not designed for sustainability. They should have been designed for reuse—the ability to recycle is not enough. Not everyone makes an effort to recycle. According to NBC News on October 4, 2007, there are over 3 billion such devices and only 15% are recycled. About 200 million tons, with mercury in the monitors and lead in the solder, wind up in landfills and often in drinking water.

Most designers are likely to miss many of the requirements noted above. This knowledge is not new. It can be included by inviting experts in these areas to brainstorm. There is no mechanism for customers to specify all of these. Suppliers that want to do productive work will teach customers how to develop good requirements as a team member. This makes the customer understand what needs to be in the contract. The point here is that if we have to fix many mistakes later (expensively), we cannot be proud of reliability, as craftsmen once were.

Paradigm 3: Measure Reliability by Life-Cycle Costs

It is wrong to measure reliability in terms of failure rates alone. Such a negative index with unknown impact does not get much attention from management, except when there is a crisis. It is the cost of failures that is important. It should be measured by reduction in life-cycle costs. The fewer the failures, the lower is the life-cycle cost. The costs should be measured over the expected life. They are not just warranty costs; they include the cost of downtime, repairs, logistics, human errors, and product liability. When I was in charge of the reliability of the Baltimore Rapid Transit Train system design, the reliability performance was measured in terms of cost per track mile. Similarly, at Baltimore Gas & Electric, reliability is measured in terms of cost per circuit mile. Smart customers look for only one performance feature: the life-cycle cost per unit of use. Those who approve the specification should concentrate on this measure. Reliability must result in cheaper, faster, and better products.

Paradigm 4: Design for Twice the Life

Why twice the life? The simple answer is that it is the fundamental taught in Engineering 101, which seems to have been forgotten. Remember 100% design margin? Second, it is cheaper than designing for one life if we measure reliability by the life-cycle cost savings. A division of Eaton Corporation requires twice-the-life at 500% return on investment [3]. It actually turns the situation into a positive cash flow, since there is nothing to be monitored if the failures occur beyond the first life. The 50% failure rate is now shifted to the second life, when the product is going to be obsolete. Engineers try to design transmission components without increasing the size or weight, using alternative means such as heat treating in a different way or eliminating joints in the assemblies. Occasionally, they may increase the size by a very minor amount, such as on wires or connectors, to expedite the solution. This is acceptable as long as the return on investment is at least 500%.

Another reason for twice the life is the need to avoid engineering changes, which seems to be obvious. Imagine a bridge designed for 20-ton trucks and a 30-year life. It may have no problems in the beginning. But the bridge degrades over time. After 10 years it may not be strong enough to take even 15 tons, and it is very likely to collapse. If it had been designed for twice the load (for 40 tons) or for a 60-year life, it should not fail at all during 30 years. It should be noted that designing for twice the load also results in twice the life most of the time, but one must still use some engineering judgment. This is similar to a 100% design margin. For the same reason, the electronic components in the aerospace industry are derated 50%. In one assembly the load-bearing capability was more than doubled by using a cheaper round key instead of a rectangular key. The round key has practically no stress concentration points. In another design, twice the life as well as twice the load capability were achieved by molding two parts as a single piece, preventing stresses at the joint. The cost was lower because no assembly was required, there were fewer part numbers in the inventory, no failures, and no downtime for customers.

What if we cannot design for twice the life? There are times when we cannot think of a proper solution for twice the life. Then one can go to other options, such as:

- Providing redundancy on the weakest links, such as bolts, corroded joints, cables, and weak components.
- Designing to fail safely such that no one is injured. For automobiles a safe mode can be that the car can switch to a degraded performance with enough time left to reach home or a repair facility.
- Designing-in early prognostics-type warnings so that the user still has sufficient time to correct the situation—before failure occurs. One of the purposes of prognostics is to predict the remaining life.

Paradigm 5: Safety-Critical Components Should Be Designed for Four Lives

The rule of thumb in aerospace for safety-related components is to design for four times the life. A U.S. Navy policy (NAVAIR) is to design safety-critical components for four times the life and conduct a test for a minimum of twice the life. The expected life should include future increases in load. Many airlines use their aircraft beyond the design life by performing more maintenance. This indirectly exposes many components to work beyond the normal one life. This is the main reason for designing for four times the life, to maintain 100% design margin all the time. Similarly, many consumers drive cars far beyond the expected 10-year life.

We should also design for peak loads, not the usual mean load. When a high-voltage cable used in power lines broke easily, engineers could not duplicate the failure with average loads. When they applied the peak loads, they could.

Designing for four times the life does not mean overdesigning. It is the art of choosing the right concept. If the attention is placed on innovation rather than marginal improvements, engineers can design for multiple lives with little or no investment, as shown earlier by several examples. They must encourage themselves to think differently rather than latching on to outdated traditional methods of increasing the size or weight. Engineers who talk of costs when solving problems usually block out creativity. They draw the boundary around the solution. Their first thought is to increase the size or weight to design for high loads. This is very common defective thinking. This is where the universities need to be more knowledgeable. We need to balance logic with creativity and should still be able to show a high return on investment.

Paradigm 6: Learn to Alter the Paradox of Cost and Performance into a Win–Win Situation

Most engineers are of the opinion that high reliability costs more. World-class organizations embrace the paradox of increasing reliability and lowering costs simultaneously. Trade-off between reliability and cost is not always necessary. Toyota has mastered this paradigm, where high reliability and lower life-cycle costs are a way of life. Toyota has learned over the years that preventing failures is always cheaper than fixing them if the failure prevention process starts early in the design. If we capture the potential failures during the requirements analysis, we can include design for reliability without making wasteful engineering changes later. Similarly, during detailed design reviews, such tools as design failure modes, effects, and criticality analysis (FMECA), process FMECA, and fault tree analysis, if used early, can help us discover many missing, vague, and incomplete requirements. Engineering changes are the biggest source of waste in organizations, because most of them can be prevented. Here are some examples of achieving high reliability with very little or no investment. Since high reliability reduces life-cycle costs, the insignificant amount of investment does not negatively affect the win–win scenario.

Example 1

A company in Brazil had designed a large warning light bulb on a control console, with a plastic cover to reduce glare. They told me that they tried all kinds of plastics for the cap but that all of them melted after a few months. Someone suggested using a glass cover. We received the usual stupid answer: "Glass will cost three times as much as plastic. The cost of the product will be high." The bad part is that many engineers look only at the cost of the component and completely ignore the cost of losing customers and the warranty costs to the employer. They are unaware that the cost of getting a new customer is at least five times the cost of retaining a current customer. When the team calculated the life-cycle costs of plastic versus the glass cap, the return on investment (ROI) turned out to be 300% in favor of the glass material. The author requested them to put a hold on the solution because we had agreed on an ROI goal of at least 500%. The author advised the entire team to take long showers for three weeks in the hope that someone would come up with a better idea. Why? Because when you take a long shower, your brain is calmed. In this state it is able to use over 1000 billion neurons that you have never used.

It so happened that the present author (the facilitator) was the one taking the long shower. Suddenly I began to feel that the engineers were giving me a snow job! They said that they tried all the plastics and they all melted. This could not be true. There are fundamentally two types of plastics: thermoplastics, which melt with heat, and thermoset plastics, which harden with the heat. I sent them an e-mail suggesting that they try thermoset plastic. It worked. They could not melt it, no matter how much heat they put in. They sent a nice e-mail: "Thanks for the research you did for us." The cost of the new plastic was almost the same. Zero investment. One hundredfold life. One million percent ROI!

Example 2

The original European jet aircraft Comets were cracking around the windows. They were taken out of service for two years. The engineers, as usual, started to design thicker fuselage walls and proposed an enormous cost increase. Then someone suggested examining the failures and discovered that all the failures were around the corners of the widows. He suggested increasing the radius at the corners. Problem solved quickly, with hardly any investment. The ROI was least 100,000% if you consider the ratio of the cost of thickening the fuselage and the investment in changing the radius on the corners of the windows.

Example 3

At a General Motors facility, the headlamps were failing after about 1000 hours of use. The supplier was going to raise the price 100% to design for twice the life. An engineer turned the filament in the headlamp 90° to avoid harmful vibration and the life increased at least sixfold. Practically zero investment.

Example 4

A dent in a Caterpillar tractor spring was causing premature breakdowns. The reason for the dent was that the spring under the tractor occasionally hit rocks on the ground. The engineers reduced the diameter of the spring such that it wouldn't hit rocks and replaced it with a tougher spring. With a very small investment they got a better than 10,000% ROI.

Paradigm 7: Design to Avoid Latent Manufacturing Flaws

We can design for reliability as much as we want, but if manufacturing processes are subject to operator error and to wide swings in variability, a good design is bound to have premature failures. We need to identify manufacturing features such as the correct torque for fasteners, vulnerability to installing components backward, or vulnerability to using the wrong components. These features could be certain dimensions, alignment, proper fit of mating parts, property of a lubricant, workmanship, and so on. A product should be designed to avoid such vulnerabilities or should be testable during manufacturing to detect abnormalities. For lack of current terminology, we can call it *design to avoid latent manufacturing flaws*.

Let's look at an example of designing to reduce vulnerability to manufacturing variations. A new motorcycle design involved over 50 different fasteners. Following process FMEA, the production operators discovered that a separate torque was required for each fastener joint. They approached design engineers to ask if they could choose about 20 different fasteners instead of 50. This would allows them to concentrate on fewer fasteners and fewer fastening standards. Engineers were flabbergasted: Such advice coming from the hourly workers was an aha! moment for them. They standardized on a few fasteners.

Another example is from Delco Electronics (now Delphi). A plastic panel required that a plating process have a conductive surface. The plating had been peeling off in two to three years and six sigma team efforts failed to control the plating durability. Someone came up with the bright idea of adding carbon particles to the plastic to make it conductive. The entire plating process was eliminated. The cost went down by 70%. The reliability of the conductivity was now 100%! A good example of over 100,000% ROI.

The secret of controlling manufacturing flaws is to identify where inspection is needed and to design the process such that no inspection is required—if such a solution is possible.

One more example may help. In this case, the process is the focus. Assume that we want to design a dinner table with four legs such that the legs must be equal. If we cut one leg at a time, we cannot get them all equal because of the variability in the cutting process. But if we take all four legs together, and cut all of them with a single cut, they will all be equal.

Paradigm 8: Design for Prognostics Health Monitoring

In complex systems such as telecommunications and fly-by-wire systems, most system failures are not from component failures. They are from very complex interactions and sneak circuits. Failure rates are very difficult to predict. The sudden acceleration experienced by Audi 5000 users during the 1980s was a result of a software sneak failure. A bit in the integrated circuit register got stuck at zero value, which rapidly increased the speed when the gear was engaged in reverse mode. One way to prevent system failures is to monitor the health of critical features such as "stuck at" faults, critical functions, and critical inputs to the system. A possible solution is to develop a software program to determine prognostics, diagnostics, and possible fallback modes.

The following data on a major airline, announced at a Federal Aeronautics Administration (FAA) National Aeronautics and Space Administration (NASA) workshop [4] shows the extent of unpredicted failures:

- Problems reported confidentially by airline employees: about 13,000
- Number actually in airline files: about 2%, or 260
- Number known to the FAA: about 1%, or 130

The sneak failures are more likely to be in embedded software, where it is impractical to do a thorough analysis. Frequently, the software requirements are faulty because they are not derived completely from the system requirements. Peter Neumann, a computer scientist at SRI International, highlights the nature of damage from software defects in the last 15 years [5]:

- Wrecked a European satellite launch
- Delayed the opening of the new Denver airport by one year

- Destroyed a NASA Mars mission
- Induced a U.S. Navy ship to destroy an airliner
- Shut down ambulance systems in London, leading to several deaths

To counter such risks, we need an early warning, early enough to prevent a major mishap. This tool is prognostics health monitoring. It consists of tracking all the possible unusual events, such as signal rates, the quality of the inputs to the system, or unexpected outputs from the system, and designing in intelligence to detect unusual system behavior. The intelligence may consist of measuring important features and making a decision as to their impact. For example, a sensor input occasionally occurs after 30 milliseconds instead of 20 milliseconds as the timing requirement states. The question is: Is this an indication of a disaster? If so, the sensor calibration may be required before the failure manifests as a mishap.

SUMMARY

In summary we can say that we need to define functions correctly. We need to design not to fail, and we need to implement all the paradigms covered in this chapter, including designing to avoid manufacturing problems. Once I was at a company meeting where the customers were asked to describe the warranty they would wish to have. One of them said (and others agreed): No warranty is the best warranty. Very few understood the paradox—the best warranty would be one that would never experience a claim. In other words, the customers wanted a failure-free design for reliability.

REFERENCES

[1] Kuo, W., Compatibility and simplicity: the fundamentals of reliability, *IEEE Trans. Reliab.*, vol. 56, Dec. 2007.

[2] Raheja, D. G., *Product Assurance Technologies*, Design for Competitiveness, Inc., 2002.

[3] Raheja, D. G., and Allocco M., *Assurance Technologies Principles and Practices: A Product, Process, and System Safety Perspective*, 2nd ed., Wiley, Hoboken, NJ, 2006, Appendix.

[4] Farrow, D. R., presented at the Fifth International Workshop on Risk Analysis and Performance Measurement in Aviation, sponsored by FAA and NASA, Baltimore, Aug. 19–21, 2003.

[5] Mann, C. C., Why software is so bad, *Technol. Rev.*, July–Aug. 2002.

Chapter 2

Reliability Design Tools

Joseph A. Childs

INTRODUCTION

The importance of designing reliability into a product was the focus of Chapter 1. As technology continues to advance, products continue to increase in complexity. Their ability to perform when needed and to last longer are becoming increasingly important. Similarly, it is becoming more and more critical to be able to predict failure occurrences for today's products more effectively and more thoroughly. This means that reliability engineers must be increasingly effective at understanding what is at stake, assessing reliability, and assuring that product reliability maturity is at the level required. To assure this effectiveness, tools have been developed in the reliability engineering discipline. This chapter is a summary of such tools that exist in all aspects of a product's life: from invention, design, production, and testing, to its use and end of life.

The automation of reliability methods into tools is important for the repeatability of the process and results, for value-added benefits in terms of cost savings during the application of design analysis methods, and for achieving desired results faster, improving design cycle times. As design processes evolve, the tools should evolve. Innovation in the current electrical and mechanical design tool suite should include interfacing to the current design reliability tool suite.

Design for Reliability, First Edition. Edited by Dev Raheja, Louis J. Gullo.

Reliability Tools in the Life Cycles of Products

One important thing about reliability engineering as a discipline is that it is involved in all parts of a product's life: from product inception, its manufacture and use, to its end of life. This is because reliability is an intrinsic part of a product's essence, whether it is a "throwaway" coffee cup or a sophisticated spacecraft intended to last 10 years in outer space. As an intrinsic parameter, it must be taken into account in the definition, design, building, test, and use (and abuse) of the product. For each program phase, tools have been devised to enable engineers to gain insight into the requirements and status of reliability. Figure 1 provides a generalized flow, representing any product's life cycle and how reliability mirrors those phases throughout a development program. Figure 2 notes key activities and events throughout a product's life cycle.

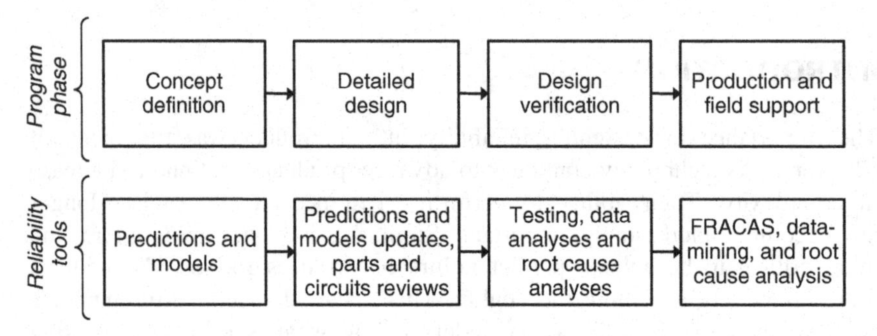

Figure 1 Reliability involvement in program and product life.

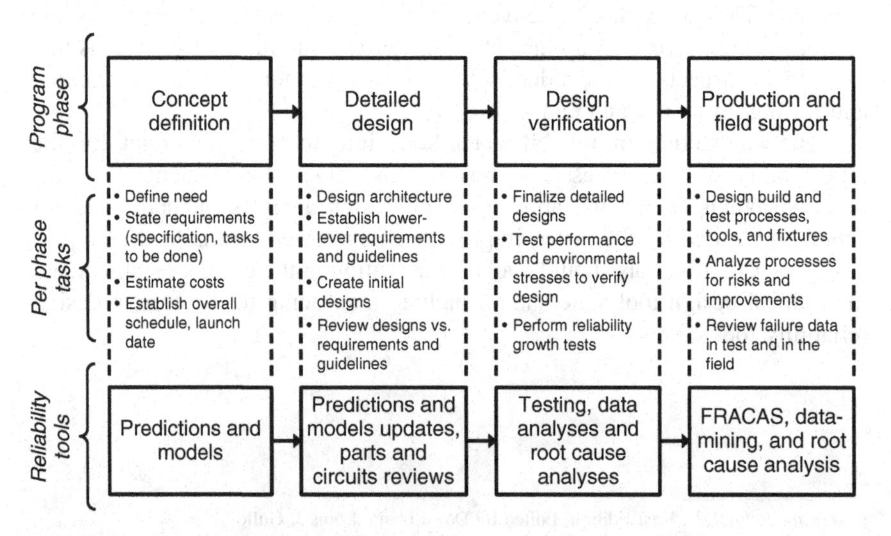

Figure 2 Program and product life tasks.

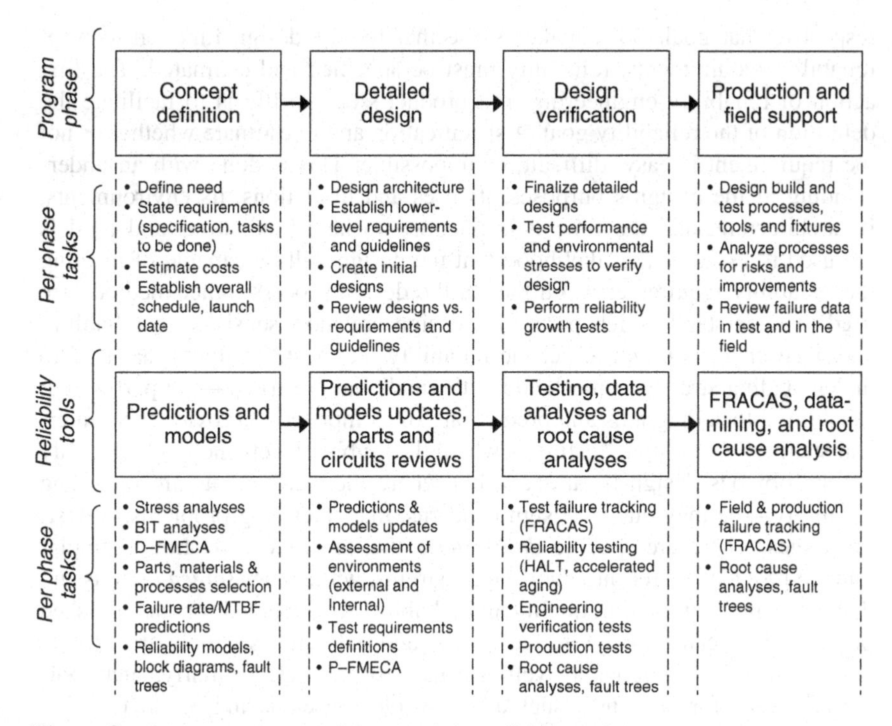

Figure 3 Program and product life tasks, tied to reliability tasks.

The reliability tools are designed to help the reliability function to assess and enhance the design so that the product is capable of meeting and exceeding its goals.

In this chapter we provide an overview of many of the tools used in the design life of a product: what they are, how they are performed, and how their results are used by the various design disciplines—reliability, electrical, mechanical, and software design, test, and manufacturing engineering. Figure 3 illustrates the reliability tools that are discussed here, when they might be used in a product's life cycle, and how these tools match the actions and events in each phase of a product's lifetime.

The Need for Tools: Understanding and Communicating Reliability Performance

Engineering

For anyone to meet a goal, at least two pieces of information are needed: what the goal is, and where the existing product or action is estimated to be with respect to that goal. So it makes sense that for the design function to meet reliability requirements, reliability must be specified and estimated. The first action of reliability engineering in a product's design life is to facilitate the definition of the reliability goal or specification and to estimate whether or not the requirement is easy, difficult, or impossible. This is done with an understanding of the design's purposes, its uses and limitations, its environments, its architecture, and the time it is intended to last. Predictions and models are used to estimate the likelihood that the design will be capable of meeting the reliability requirements. Further in the design process, other methods are used to understand which parts of a design are most sensitive to reliability and what changes might affect the reliability the most. Analyses that lead to understanding stresses—mechanical, thermal, and electrical—are performed; reliability block diagrams and predictions are completed to provide the numerical status of the design. Testing is where the "rubber meets the road" not only for the obvious design parameters, but where the weak points are regarding reliability—be they due to design concerns, manufacturing problems, or testing issues. There are specific *design tests* that are used in testing exemplar units to provide insight into the design margins, while other *screen tests* assure that a minimum amount of quality and reliability are included for each finished product. Root-cause analyses, fault trees, and data analysis tools are used to detect test result concerns, as well as causes for failures. Similarly, such tools are also used for defining issues and possible causes during production and use in the field. Often, designs must undergo modifications due to changes in requirements or obsolescence of parts. When this occurs, many of the tools used in early product life are revisited and updated.

Management

The reliability status must be provided to management to assure that the performance, schedule, and cost risks are understood with respect to the goals. Clearly, besides the obvious cost and schedule losses due to missing a reliability requirement, there are many subtle but equally important losses: customer dissatisfaction, loss of brand quality, and ultimately, loss of repeat business. Another key consideration is the possibility of safety issues due to potential accidents and mishaps. Therefore, it is important that the ability to meet goals, as well as potential technical problems related to failures, be communicated to management in an objective and useful manner. For the tools to be effective, they must facilitate fulfillment of this need.

Customers

The real winner in designing and building a product that meets specifications with margin is the customer. This means that the supplier designed and built the product as intended, and that the product lasts. In some cases, such as for space or military customers, the customer representatives are involved directly in the development process. As such, they are informed as to the reliability status throughout the design process. But even commercial customers are swayed by studies of quality and reliability of products, so that it is always to the producer's benefit to know what the product reliability is and how it will be received by the public. The most important test of any product is how well the product stands up to customer expectations. The reliability tools help to define reliability goals and logically assess progress toward those goals. The benefit to using such tools is to provide a means of communicating such assessments: to management, to design engineers, and to customers.

RELIABILITY TOOLS

Early Program Phase (Concept Definition)

Reliability Predictions

One of the most useful, but poorly understood, tools available to the reliability engineer is that of reliability prediction. A prediction can help us to understand the risks involved in setting reliability goals—sort of a "sanity check" for comparison to an expected capability. It can point to which portions of a product could be most sensitive to reliability concerns. It is useful in gaining insights into how part quality, application stresses, and environments can affect the overall reliability performance. Using reliability block diagrams (described below) can help one to understand the need for redundancy or partial redundancy ("m-out-of-n" redundancy).

The part about prediction that has the potential to be misunderstood is that they are not accurate. George E. P. Box, a well-known industrial statistician, is attributed as the author of the saying: "Essentially, all models are wrong, but some are useful." This is especially true for reliability predictions. The great majority of predictions are performed by reliability practitioners using industrial or military standards, such as the U.S. military's MIL-HDBK-217 [1], Bellcore TR-332 [2], and Telcordia SR-332 [3]. The intent of each of these documents is to provide a standard method of calculating the failure rate of components used in specific environments with given stresses and quality ratings. In general, to use them you must assume that the equipment has a constant failure rate. For electronic parts for which high output quality and special screening has become the norm, this is generally true. For a large number of components in an assembly or system, the total failure rate can generally be assumed to be constant. At least for assemblies with long-life components with mixed failure modes and mechanisms, the composite assembly failure rate is reasonably constant. However, the question of validity comes into play when one looks at details, such as the use of specific model types for specific parts that are provided in the standards. In many cases, average factors and model constants are used to estimate the failure rate. These "averages" can be tweaked over time by incorporating field and test failure data (known as *Bayesian techniques*). This is useful in later phases of the program, but early on, such data are available only from historical data associated with similar products.

Mechanical parts, mechanical failure modes of electronic and electrical parts, and cyclic failure mechanisms are not fully comprehended in the methodology described above. In actuality, these are typically not constant failure rate modes. In these cases, where knowledge of the failure rates is critical, the most comprehensive method for predicting product life is *physics of failure* (PoF): that is, to understand how the failure modes are caused and to use models applicable to the underlying failure strength-to-stress interactions. This PoF modeling has a much higher fidelity than use of the standard techniques and should be used wherever practicable

PoF is using the underlying science behind failure mechanisms to model each of the key possible stress–strength relationships. In many cases such models already exist, but for new technologies, applications, or packaging, such models may require development by test. Not only are the root causes of each failure mechanism modeled, but where possible, the variabilities associated with the stresses and the material properties can be modeled to gain insight into expected failure time limits. This type of analysis results in an understanding of potential causes of failure, times to failure, and possible margins. As with any of such predictive models, it is wise to validate the models with actual stress tests. This type of analysis requires comparatively high resources, and in most programs is presently used for particularly critical applications, as cited above.

The prediction results are not just numerical values, such as mean time between failures (MTBF), failure rate, or probability of success. For them to be meaningful, their calculation methodology and underlying assumptions must be communicated as well. This allows the receiver of the information to make an informed judgment as to how much weight to give those predictions versus the risks involved. In the concept phase, these are used by managers and customers to develop an idea of the product reliability capability and where reliability should receive the highest priority. They also provide input to engineers to optimize the device to withstand applications of overstresses and environmental extremes.

Reliability Models: Fault Trees and Block Diagrams

The underlying idea behind reliability block diagrams (RBDs) and predictive fault trees is to display the logic of how a failure could interrupt the intended

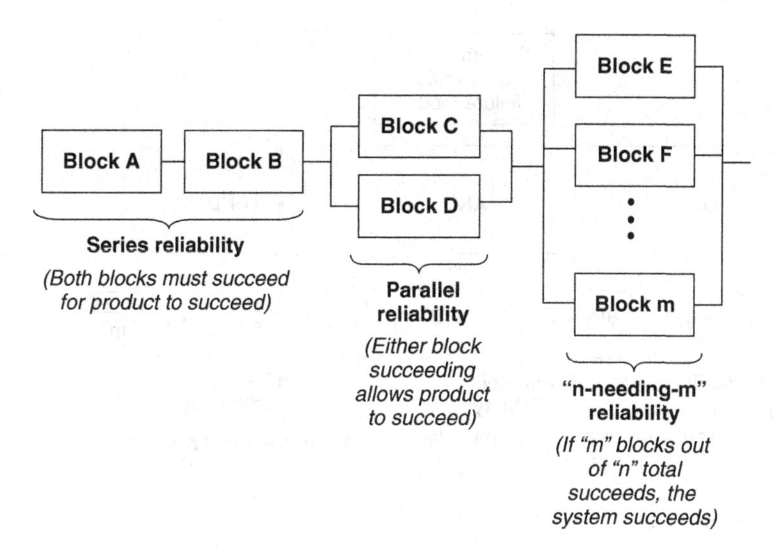

Figure 4 Reliability block diagram.

outcome or "mission." Blocks A and B in Figure 4 are "series" units in that the failure of either of these two items would result in a system failure. The probability that both of these are successful is the product of their probabilities: $R_{\text{segment}} = R_A \times R_B$. In the general case, the segment or system reliability for n blocks is

$$R_{\text{segment}} = \prod_{i=1}^{n} R_{\text{blocki}} = R^n \qquad (1)$$

The C and D devices are *in parallel* or *redundant*. If one device were to fail, the second would be sufficient to allow the system to succeed. In other words, both must fail for the system to fail, where the probability of both failing is the product of their unreliability $1-R$, or in this case, $R_{\text{segment}} = 1 - [(1 - R_C) \times (1 - R_D) \times \cdots \times R_m)]$.

Blocks E through m represent an m-out-of-n redundant configuration. In other words, this segment is still successful as long as n blocks are operating. So, for example, if there are 15 devices in the segment but the system will continue to succeed as long as there are still 12 in operation, this is a "12-out-of-15" redundant segment. In the general case, the segment or system reliability formula is

$$R_{\text{segment}} = R_{m\text{-out-of-}n} = 1 - \sum_{i=0}^{m-1} \left\{ \frac{n!}{i!(n-i)!} [R^i (1-R)^{n-i}] \right\} \qquad (2)$$

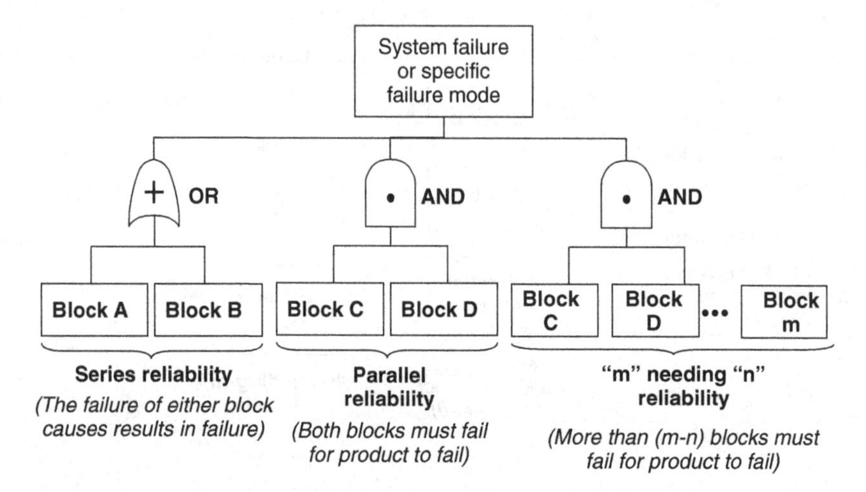

Figure 5 Fault tree.

The RBD is a picture of the paths that result in a success, while fault tree logic provides a similar logic picture for a failure or a specific failure mode. In both cases, the results are used to develop and communicate the Boolean logic used to provide a probabilistic model for reliability—the probability of success.

The fault tree (Figure 5) is an analysis tool that is like the "mirror image" of a RBD: Where the RBD tool helps to calculate the probability of success, fault tree logic assists the user to calculate the probability of failure. The elements, then, are blocks that provide causes of failure due to lower-level failures or events, and if done properly, to the root causes. One other key element of fault trees is that they generally include Boolean algebra symbols to show series and parallel logic. When completed, the probabilities of the lowest level of causes or failures can be used to develop an equation to provide the probability of failure. The fault on which the fault tree is based may be a system failure, or it could be the failure of a portion of the system or a specific failure mode of the system or system component.

Either of the two tools—RBDs or fault trees—can serve to provide a picture of the logic underlying the probability calculations regarding the probability of the product lasting through a particular period of time, whether that be a specific use or mission phase or its entire expected life. As with many of the reliability tools discussed here, it also serves to provide those using the models with a means of focusing on the key drivers to those numbers and clues as to how the reliability might be improved.

This type of analysis is very useful in trading off the use of redundancy versus single series items. Even if the reliability is a comparatively high number, say 95%, such ideas may need to be explored: for example, if a system requires a reliability of 90%, but a single item X has a reliability of 80% while the rest of the system reliability is 95%. The system reliability would be the product of these two probabilities, or 76%.

However, if a second 80%-reliability item X were added so that only one of the two X's was required, the resulting reliability of the two X's would be 96%. Then the system's reliability would be 96% $\times$ 95%, or 91.2%. If more margin were needed, additional redundant units would increase the reliability further. However, there are very important factors that must be considered beyond just the numbers. Redundancy adds complexity (e.g., added fault detection and switching must be incorporated). Cost, weight, and space are additional program concerns that must be taken into account.

Reliability Allocations

One tool that is widely used is that of allocation of reliability. It is usually included in a reliability program as a means of providing specifications or goals at multiple levels in the design. So if a system has an MTBF requirement of x, four equal system components or modules would have an allocation of $4x$. This stems from the constant-failure-rate assumption: If the example system MTBF is x, its failure rate is $1/x$ and its four equal modules would each have a failure rate equal to $1/4x$. The MTBF of each module would then be the inverse of $1/4x$, or $4x$. Similarly, then, one can use this idea for different modules with different failure rates by apportioning the failure rates so that their sums are less than or equal to the system requirement.

Of course, this does not really work for nonconstant failure rates, where the addition of failure rates is not a valid process. In this case, the known models can be incorporated into RBD or fault tree probabilities and allocated appropriately using probabilistic models. It is also difficult to allocate failure rates among modules for which there is little or no information (i.e., in the concept phase or for new designs or technologies). In this case rough estimates of complexity or reliability can be used to help divide the reliability "budget" to provide lower-level specifications.

The intent of the allocation is to provide a tool for achieving a higher-reliability goal by budgeting the requirement to the lower-level designs (in many cases, multiple levels) to provide goals for the design engineers and their leaders.

Parts, Materials, and Process Selection

Whether designing a product from a blank sheet of paper or when developing or modifying an existing design, one of the key decisions that designers must make is which parts to use for specific applications. Of course, the most

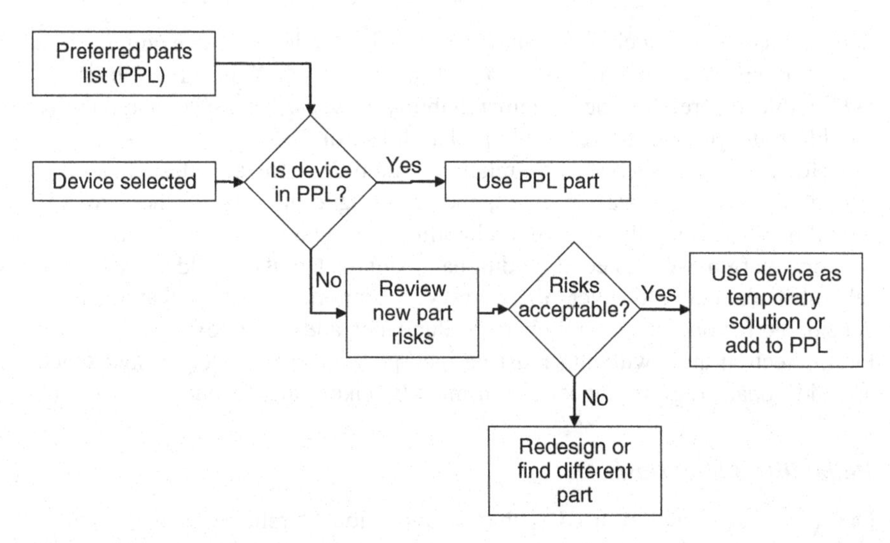

Figure 6 Preferred parts flow.

important factor is performance. The part must meet specific performance parameters in order to meet the requirements of the design. But the next tier of factors includes cost, availability, life cycle, commonality, quality, and reliability.

A preferred parts list helps to manage this decision-making process. The point is to provide straightforward access to parts that have been reviewed or used from known suppliers with positive quality and reliability credentials. This helps to avoid "surprises" such as a new supplier or part type that fails in unexpected ways or due to unknown causes. It can also help to assure that common parts are used within a product's development: useful in managing cost, schedule, and technical issues with a part or supplier. In some cases the supplier or part would perform as needed, but due to new materials or configurations, the use of such parts can represent new technologies in the manufacturing arena. These types of issues frequently result in reliability risks. So use of a part that is not in the parts, materials, or process list must go through a special review. Figure 6 is a flow diagram that shows how this list would be implemented in a process.

This process provides a framework for the designer and the components engineer to standardize parts or to make an informed decision for specific instances. This is not to say that all new technology or suppliers cannot or should not be used. It is an "insurance policy" that assures that such selections are fully reviewed and the risks delineated.

Stress Analyses and Design Guides

Design guides help to serve as a means to communicate lessons learned concerning reliability issues. In this context, the term *derating* is often used to mean applying a part that is better than its requirements. For example, if a capacitor is expected to work in a circuit at 5 V dc, then selecting a 10-V dc capacitor will add life to that device. This is a 50% stress ratio, sometimes referred to as *50% derating*. Derating is important, because often the stress ratio on a device is related exponentially to the device's life expectancy. Choosing a low stress ratio often adds little or nothing to cost or schedule risks, so using them is clearly a benefit. The guidelines can also be used to caution against certain types of parts or circuit designs. This can be of great value in avoiding risks before they are designed-in and require modification.

Stress analysis is analysis of the design to compare the application stresses to the rated value and the derating recommended in the guidelines. It is typically performed during the design process to determine the stresses—voltage for capacitors, current or power for resistors and inductive devices, and junction temperature and power for active devices and integrated circuits—so that the guidelines can be followed. These choices also provide input into the reliability predictions and will affect that assessment as well.

When it is not possible to follow these guidelines due to the lack of availability of a part or special performance needs, these exceptions should be reviewed by peers and technical management to assure that informed decisions are made and that the requirements are properly balanced in the design before proceeding to the building and test phases.

Design Failure Modes, Effects, and Criticality Analysis

There are a variety of approaches to design failure modes, effects, and criticality analysis (D-FMECA). One of those approaches is offered in Chapter 5. A different approach to FMECA is shown in Chapter 6, but with an emphasis on process rather than design. MIL-STD-1629 provides a separate approach, often used in the military weapons design process. Regardless of which type is used, there are common uses and goals for FMECA. This type of analysis considers the possible ways in which parts or system components can fail (modes), the impact of such failures on the overall product and its use or mission (severity) and the likelihood of their occurring (frequency or rate), and the modes' ability to be detected. Another key part of these analyses is the determination of an importance or ranking of each failure mode considered.

The output of D-FMECA allows the designer to learn what failures are important and why, and to document the findings and their rationales. Each analysis should result in serious consideration of follow-up actions. Upon completion of FMECA, the entries should be ranked to find those that have the highest risk: those that can cause severe outcomes, are likely to occur, and can occur without warning or the ability to locate the problem. The highest-ranked entries should then be reviewed so that proactive responses can be made. That is, review of the architecture for possible redundancy, selection of improved quality parts, derating, self-test considerations, and even how the device in question is installed and handled. This is not to say that FMECA is used only to make design or other changes, but it serves as a pointer to issues for consideration. Other uses of FMECA outputs are to feed RBD and maintainability calculations and to support detailed testability or built-in test (BIT) analyses. However, the real leverage from D-FMECA is to provide a ranked picture of where the highest risk is in the design for further action.

BIT Test Definition and Effectiveness Analysis

Testability analysis is a multidisciplinary engineering effort involving the design and review of circuitry that either tests itself or can be tested within the product being designed. This capability of testing itself is called *built-in testing* (BIT). However, to take advantage of such features it is necessary to translate the intent of the BIT, as well as how the circuitry works, into a language that software engineers can use. That is, for a system or product to test itself, its processor or control sequencing function must provide the proper stimuli and interpretation of the responses. This allows the product to provide feedback with respect to a “good” or “bad” result or a level over and/or under defined thresholds. When performing this translation, the testability engineer must also take into account variability in the design due to part-to-part differences, changes in conditions such as temperature, and manufacturing or material variations.

When a design is complete, the circuitry is tested with and without faults inserted to assure that proper tests and test limits have been incorporated. Also, the testability engineer can estimate the effectiveness of BIT by analyzing which parts and part failure modes can be detected and can compare the failure rate of the detectable faults to the total product failure rate. This ratio of detectable failure rate to total failure rate is the BIT Effectiveness or probability of fault detection.

Fault isolation is another key parameter when a product is to be maintained. For example, the “check engine” light in a car can tell the driver that something is wrong. This is fault detection. However, that light, combined with “oil low,” helps us to understand the problem both operationally and for “repair”. In other words, it helps to know what to do immediately and then how to go about finding out what the actual problem is. This reduces the problem to one or a few possible causes. The fewer the items left to explore, the better the isolation is.

The probability of fault isolation is also a failure-rate-weighted average of the items that can be isolated given detection. The isolation requirements can be defined in terms of a single component or groups of x components. That is, a probability of fault isolation might be specified as an 85% probability of isolating a fault to a single unit (board, box, replaceable element), or it might be specified as a 95% probability of isolating a fault to five or fewer elements.

Isolation analyses are performed to provide a measure of goodness regarding the testability design of a product. Like many other such measures, the importance of this effort depends on the product's intended use. A simple throwaway toy, for example, is not a likely place where one might spend resources designing-in testability. On the other hand, for a device that is maintained, such as a car, a piece of manufacturing equipment, or a weapon system, such design measures are critical. In these cases, communication of testability findings with management, design/test engineering, and the customer or user, is vital to the success of a product.

Detailed Design Phase

Continued Stress Analysis

Often, a design is in a state of flux during the concept phase of a program. Therefore, it makes sense to continue to update the stress analyses as the design matures and to compare the results to the guidelines. This also allows the reliability analyses to be abreast of the design changes. The stress analysis update provides design engineering with an objective view of the applications and engineering management with an overview of potential reliability issues, which facilitates informed decision making. If followed, this helps to maximize the reliability of the final design before it is manufactured and tested.

Process FMECA

Process FMECA (P-FMECA) provides much the same type of information as that of D-FMECA, but with an important difference. It focuses on the producibility of the product: what is involved in materials handling, assembly, adjustments or calibration steps, and testing. This is important to the design of a product because the ease of manufacture often affects reliability greatly. For example, if new packaging is required, special installation methods and tools may need to be designed to alleviate specific issues. Sometimes, the design must change to accommodate a manufacturing issue that improves quality or cost issues later in the product life cycle.

Continued Use of Tools

It is important to note that the tools covered in the "early program" are still useful in later stages. Their validity and accuracy can often be improved greatly once the design details become more developed. It is in the interest of the program to gain visibility into the reliability issues as the design develops so that changes can be known and addressed.

Design Verification Phase

Failure Reporting and Corrective Action System

The failure reporting and corrective action system (FRACAS) is one of the most useful tools available to reliability engineer or to anyone who needs to understand the product reliability or how it might be improved. It is a method of collecting failure data of a product such that it facilitates the ability to look at different "pictures" of the information to show the failure history.

The idea is to provide a receptacle—usually a database—for specific factors surrounding a specific test anomaly or set of symptoms, using a standard set of inputs. These inputs then can be used to select and sort specific aspects of failure events so that the data can be prioritized using Pareto analysis or other ways to show quantities versus specific parameters. The parameters can be such factors in a report as part type, supplier, date codes, circuit cards or modules, serial numbers, date of events, or test type. This allows the user to detect failure trends, review failure histories, and gain insight for investigating the causes of failures.

The intent of FRACAS is to provide a feedback loop for failures. As tests are performed and events occur, more can be learned about causes of failure, allowing corrective actions or design improvements to be implemented. A failure review board (FRB) includes a mix of key disciplines (design, manufacturing, systems, quality, and reliability engineering) as well as leadership (technical or managerial leads). This assures a well-rounded discussion of specific issues associated with failures, and optimal actions taken with respect to root-cause analysis and corrective follow-up. Ultimately, management is responsible for the decision making but can do so with added visibility from FRACAS results.

Root-Cause Analysis and Fault Trees/Ishikawa Fishbone Charts

The underlying idea in the use of FRACAS and tying test and field failures to a means of discovering and improving system weakness is largely dependent on finding the cause of failures or failure trends. Often, failures are analyzed to the "immediate cause," but without delving deeper, the underlying cause is

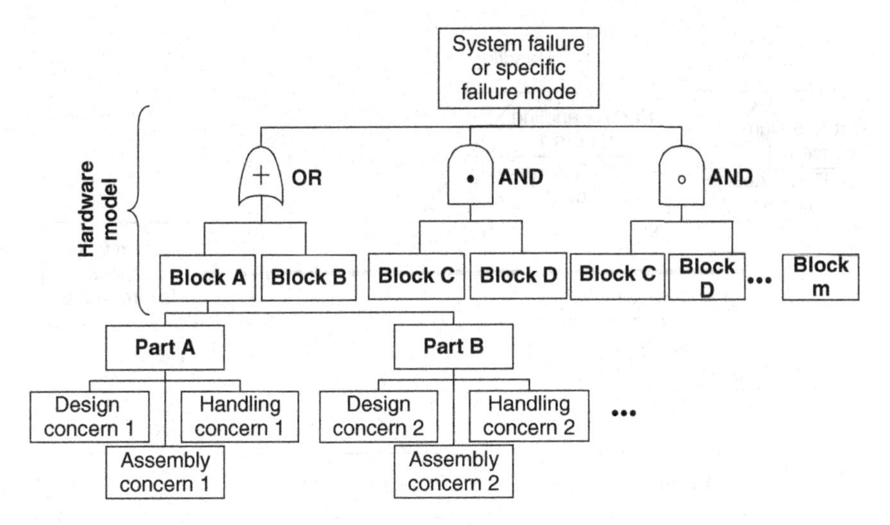

Figure 7 Fault tree.

not addressed. This can result in "repeats" or reappearance of the same failure modes. In other words, for robust reliability growth to occur, it is imperative to uncover and address root causes. Two tools that can assist in that effort are fault trees (Figure 7) and Ishikawa fishbone charts.

Fault trees in this application are used for a purpose other than as a prediction tool. In this case the "fault" is an unwanted failure or event. The underlying logic is still the same: to iteratively assess the cause at multiple levels to delve into causes of the causes until a root cause is found. In this context, a *root cause* is *an actionable process step or steps that, if addressed, would eliminate the possibility of the top event*.

This is a different application than that used in predictions, so it is used differently:

- As the causes are uncovered, not only is the failed part or assembly considered but also what process or actions caused or contributed to the likelihood of the event.
- There is little or no use of reliability predictions or probabilities to calculate the probability of failure. The idea is to gain an understanding of what could have caused the occurrence and address such causes or contributors in actual practice.
- Boolean gates are often not used except for the first hardware configuration modeling. Usually, the processes involved are not redundant and therefore do not need to consider AND and OR gates. If it is found that redundancy does need to be considered, they can easily be added.

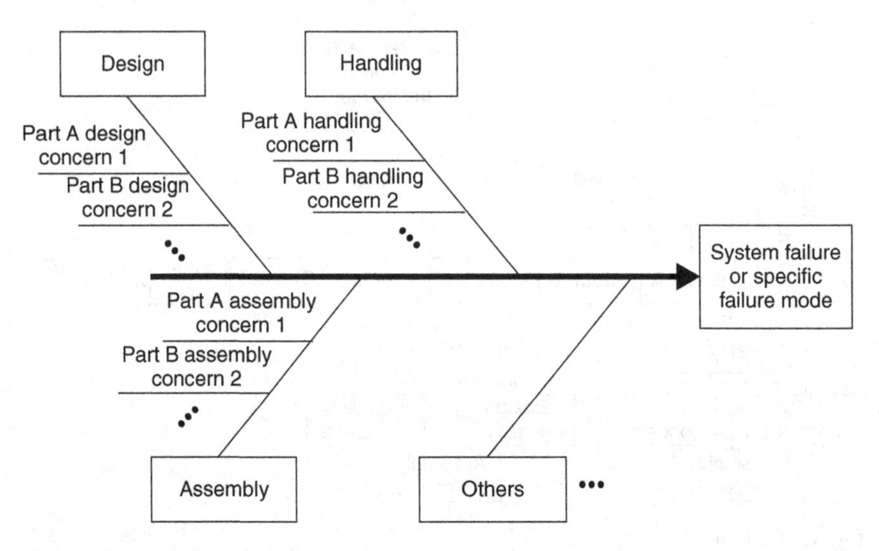

Figure 8 Ishikawa fishbone diagram.

- Processes can be modeled at the top level as well. However, experience shows that the event first requires troubleshooting to some level before process steps can be considered to the depth needed.
- The output of this process should be the cause or causes addressed to develop reliability improvements (i.e,, reliability growth). This is not the same as probability of failure, where actual events have not occurred, but it, too, can help focus on high-probability issues. Both can be used to help drive improvement action.

An Ishikawa fishbone chart (Figure 8) is almost the same as a fault tree but has a different format. It can be designed to help assure that the user considers all the important processes involved. Some typical areas considered in either a fault tree or fishbone diagram to find root causes are:

- Assembly [documentation, tool calibration, error-proofing, and electrostatic discharge (ESD) prevention]
- Human error (task loading, training, and distractions)
- Design (materials, application, environmental stresses, and engineering review processes)
- Handling (point-to-point movement, inspection handling, and protection)
- Test (test equipment, calibration, and ESD prevention)

TEST DATA ANALYSIS

Life Testing and Acceleration Factor

Specific failure modes are often tied to a component life. *Life* in this context is different from the constant failure rate discussed earlier. It represents the part of a component's life cycle in which the failure rate is *increasing*. Reliability life tests are designed to provide insight into when in a product's life cycle the failure rate begins to increase with time or use. Such tests focus on specific failure mechanisms to facilitate the generation of *acceleration factors*, which allow devices to be tested at very high stresses to allow insight into how that same device would behave in actual lower stresses without testing very high quantities of units over very long periods of time.

Life tests are performed by applying a stress or stresses to a component and tracking the time to failure of a portion or all of a sample set. If sufficient samples are tested to failure, the failure time versus failure quantity can be analyzed statistically to develop a *failure probability distribution function* (PDF), a statistical model of the probability of failure versus time. If the life tests are repeated at multiple stress levels, a different type of model, a *stress model*, can be developed that aligns the time to failure for a specific percentage of failures (a *percentile*) versus stress.

This stress model can simply be extrapolated if it can be made into a linear equation. For example, the Arrhenius model is largely tied to chemically dependent failure mechanisms such as corrosion or chemical reactions. It reflects the proportional relationship between time to failure (t) and the exponential of the inverse of temperature (T):

$$t_{\text{test}} \propto e^{\varepsilon/T_{\text{test}}} \tag{3}$$

where ε is a constant called the activation energy, which is dependent on the materials involved and T is the absolute temperature in kelvin. Assuming that the PDF does not change over the stress (in this case, temperature), dividing both sides by the same equation for actual stresses yields

$$\frac{t_{\text{test}}}{t_{\text{actual}}} \propto e^{1/T_{\text{test}}-1/T_{\text{actual}}} \tag{4}$$

This ratio is the acceleration factor, α:

$$\alpha = \frac{t_{\text{test}}}{t_{\text{actual}}} \tag{5}$$

The α factor is used to perform tests at higher stresses (in this case, temperatures) that allow direct correlation between the times to failure at a high test temperature to the temperatures to be expected in actual use. As long as the temperatures are such that new failure mechanisms are not introduced, this becomes a very useful method for testing to failure a relatively small sample of items for a much shorter amount of time.

This type of analysis is very dependent on knowing the specific failure mechanism being considered and the underlying relationship of that mode to the stresses provided in the test. In practice, a product such as a circuit card is often made up of many parts, each with its own failure mode, but the main drivers are microcircuits. In this case, engineers often use the same physical relationships but assume conservative factors in the models, so the α factor is a "greater than or equal to" value. Such modeling allows engineers and managers to understand how long equipment might last in an extended exposure to certain environments and how changing those environments could affect the reliability as well.

Reliability Growth Modeling

Reliability testing is a method of test, analyze, and-fix that helps to assure that a product at the start of its development can eventually mature to meet its reliability goals. The underlying assumption is that in the concept and design phases of a program, not enough is known about a product to expect it to meet its expected reliability capabilities from the start. So various environmental tests are performed, which are similar to the use environments expected, or the failure modes and their related acceleration factors are known, so that test time can be plotted against failure rate or mean time between failures. It was observed that if these data are plotted on a log-log plot, the plot generally becomes linear over time. Using this knowledge, the data in such testing can be used to understand what the failure rate of a unit under test is at any given point in the test, and if performed properly, the testing will result in a decrease in failure rate such that the data can be extrapolated. This allows the reliability engineer to estimate how much testing is required before the failure rate goal is reached. It can even be used to judge the effectiveness of the FRACAS/FRB process. If the slope is not steep, it may be necessary to increase resources in that area. Good design programs have been known to have a slope of at least 0.3. Reliability growth modeling is an invaluable analytical tool for estimating reliability, testing requirements, and process effectiveness for technical management.

Production and Field Support

The FRACAS described previously is not limited to reliability testing. Events in any testing and field returns should be considered opportunities for reliability improvement. This includes engineering verification, qualification testing, acceptance testing, and field returns. If the FRB with a built-in root-cause analysis process is utilized, there is continued opportunity to gain insight and consider marked improvements over time. This type of reliability growth is systemic and based on actual use. The cost-to-benefit ratio is quite small for any product for which reliability is important.

Production Testing

There are many types of tests that involve reliability engineering directly or indirectly, due to the FRACAS methodology. The tests discussed here are those that are performed with the purpose of understanding product reliability.

Highly Accelerated Life Test (HALT) and Aging Tests Both HALT and aging are ways to develop insight into how a product is likely to behave over time. Such testing is generally performed on equipment that has a long expected life or where the reliability of the product is critical. Often, such requirements entail safety issues, customer costs, or warranties. In these cases, there is a good business case for performing such tests, because they can provide insights into possible future risks.

HALT entails single and combined thermal and vibration testing. The product of interest is mounted in a combined-environment chamber capable of random vibration and thermal excursions. Also, the unit under test (UUT) is connected to an interface or tester to allow it to be tested during exposure.

HALT is performed as a *step-stress test*, where the equipment is stressed to increasing levels in a sequential manner. The stresses are set to higher and higher extremes, finding the *nondestructive limit*, where the UUT ceases to operate properly but has not been destroyed. This allows insight into the design margin of the UUT, as well as the weakest points in the equipment. It is an excellent method of finding reliability issues on equipment that is expected to be used in specific environments.

Similarly, an aging test is an above-specification test. In this case, the assumption is that the stresses of temperature cycling, temperature combined with humidity, and salt-fog shorten the useful life of a device as they increase. The idea is to simulate in accelerated fashion the effects of long-term stresses:

- Temperature cycling on materials with thermal expansion differences
- Exposure to temperature and humidity on materials susceptible to humidity
- Exposure to a salt-fog environment, due to sealing or packaging issues

As with the HALT discussed above, this testing provides insight into the design margin and the weakest links in the design for potential improvements. Events and anomalies in both HALT and aging require a disciplined approach to root-cause analysis to assure that the findings are addressed properly and that the technical managers have the facts necessary to make informed decisions.

Environmental Stress Screening (ESS) and Highly Accelerated Stress Screening (HASS) Both HALT and aging are excellent methods of learning about the design of equipment over a relatively short span of time and affecting that design prior to committing the program to production. Once production begins, other types of testing can provide important intelligence on how well the product is being built and tested. If ESS is included in the manufacturing process, then, as a screen, it is performed on all the equipment being produced. It is designed to stress the product so that out-of-specification weaknesses are found, to preclude "escapes" being shipped to the field. The stresses are chosen to address possible failure modes found in a manufacturing process—modes such as loose fasteners, improper solder joints or weldments, missing hardware, defective components, and improper solder joints. Typically for electronic and mechanical devices, temperature cycling and random vibration up to specification level are used to uncover possible manufacturing defects. In other words, the *producer* performs the screening so that the *customer* doesn't need to do so *in use*.

Any findings are generally expensive, but not as expensive as returns and unsatisfied customers. This translates into a good business practice, utilizing the FRACAS/FRB methodology to assure that future such events do not occur. This not only keeps potential problems out of the customer's hands, but allows product quality and reliability to continue to improve.

HASS is an accelerated version of ESS, such that the stresses used are *beyond specification values* but well below nondestructive limits. This application of accelerated stresses assures a margin in the manufactured product and shortens the ESS cycle. HASS cycles are shorter than ESS cycles since the higher stresses do not require as many cycles or time in testing to provide equivalent effectiveness at precipitating and detecting defects and failures.

SUMMARY

This chapter covered various tools used in the reliability engineering discipline. These tools provide insight into the possible and realized reliability of a product undergoing design and use by customers. They are emphasized early in the design flow when the least is known about product capability and design but when changes are the least costly in time and money to implement. These tools require a close relationship among the design, manufacturing, and quality engineers as well as management. These tools, although not an exhaustive list of all useful tools available to reliability engineers, have some important characteristics in common:

- They provide insight into the relationship between goals and progress.
- They help to identify where the issues are and what actions are required.
- They require input from other personnel in the program, since they encompass the very basics of the design.
- They provide ready means for communicating results to those who most need the information: designers and managers.

Many of the tools described in this chapter are discussed in greater detail in other chapters. As evidenced by the information provided, if properly implemented, reliability engineering tools span the design program from concept definition to end of life.

REFERENCES

[1] *Military Handbook for the Reliability Prediction of Electronic Equipment*, MIL-HDBK-217, Revision F, Notice 2, U.S. Department of Defense, Washington, DC, Feb. 1995.

[2] *Reliability Prediction Procedure for Electronic Equipment*, Bellcore TR-332, Issue 1, Bell Communications Research (Bellcore, now Telcordia Technologies), Telcordia, Piscataway, NJ, 1999.

[3] *Telcordia Electronic Reliability Prediction*, Telcordia SR-332, Issue 3, Telcordia, Piscataway, NJ, Jan. 2011.

Chapter 3

Developing Reliable Software

Samuel Keene

INTRODUCTION AND BACKGROUND

In this chapter we discuss software reliability, how it can be measured, and how it can be improved. Software is a script written by a programmer dictating how a system will behave under foreseen operating conditions. It typically performs its mainline function flawlessly. This part of the software has been tested extensively by the time it is shipped. Typically, problems arise for software in handling program exceptions or branching conditions when these conditions were not accounted for properly in the original design. Usually, the programmer is hurrying to implement the mainline function of the software, making it difficult to consider and deal properly with operational contingencies. The reliability analyst can help expose these conditions to the developer and get them handled properly.

Software is the embodiment of a programmer's thought processes. It faithfully executes its script. This can cause problems when the program faces circumstances that were not considered properly by the author of the code. For example, a planetary probe satellite was programmed to switch over to its backup power system if it did not receive any communications from the ground for a 7-day period. It was assumed that such lack of communication would signal a disruption in the satellite communications power system. The built-in maintenance strategy was to switch over to the backup power supply. This switching algorithm was built into the software. In the second decade following launch, the ground crews inadvertently went 7 days without communicating to to the probe. The satellite followed its preprogrammed instructions

Design for Reliability, First Edition. Edited by Dev Raheja, Louis J. Gullo.

and switched the system over to backup power, even though the mainline system was still functioning. Unfortunately, the standby system was not. So the logical error led to switching the satellite into a failed state. The satellite program was just doing what it had been told to do more than 10 years earlier. Given the operational environment, a condition precedent became a system faulting error driving the satellite into a failed state.

A spokesperson for a Southeastern power company made the following statement following a power system failure that brought the power down for most of one day during the summer of 2008. That management representative said: "The [failing system] software behaved just as it was programmed to do." That is always the case, and this statement offers no consolation. The root of the problem was that the software did not deal faithfully with an unusual operation in the way that the customer would have liked. Further, this power control system had the administrative software on the same computer as the control software. This comingling of applications also proved to pose critical operational risks.

There are many everyday analogies to software faults and failures. Any transaction that we embark on that has residual risks and uncertainties has a counterpart in software reliability. Look at the potential risks in a real estate contract. This could include a flaw in the title, the house might encroach on an easement, be in a floodplain, be flawed in its construction or condition, and so on. A prudent person wants to mitigate these faults or their likelihood and build defenses against their occurrences. Similarly, a prudent programmer will think proactively in dealing with contingent conditions, thereby mitigating any adverse effects. The reliability analyst can help the programmer by "putting a second pair of eyes on the design." This second view can be enhanced by using collaborative and analysis tools to test systematically the design's completeness and ability to handle aberrant conditions. Exploring the behavior of software under stress and variation conditions can be evaluated and documented systematically using the six sigma tool suite, including quality function deployment, mind map, and RASCI (responsible, accountable, supportive, controlling, and informed responsibilities) program flow breakdown. The aforementioned collaborative tools and other six sigma tools can be found with explanations on the Web site isixsigma.com.

Managing the programming consistencies problem is exasperated in developing large programs. This is analogous to achieving a consistent script with 200 people writing a novel simultaneously. Obviously, there are many opportunities for disconnects or potential faults in this effort.

Until the 1990s, software was typically considered not to fail (i.e., its reliability was considered to be 1.0). Nothing was thought to "break" or wear out. Hardware problems dominated the reliability concerns. Software (and firmware) has now become ubiquitous in our systems and is most often seen as the dominant failure contributor to our systems. For example, the infamous Y2K bug has come and passed. This bug came about because programmers' thought horizon did not naturally extend into the new millenium. When the clock rolled over into the twenty-first century, software that thought the date was 1900 could fail disastrously. Fortunately, its impact was less than that forecasted. Problem prevention has cost about $100 billion worldwide, which is significantly less than many forecasters predicted (cf. http://www.businessweek.com/1999/02/b3611168.htm). Hopefully, programmers, or information technology (IT) people, have learned from this traumatic experience. Y2K has sensitized the general public to the greater potential impact of software failures.

Imagine what might have happened if this problem caught us unawares and took its toll. Life as we know it could have been shocked; for example, disruptions could affect grocery store scanners and cash registers. Our commerce would have been foiled by credit cards not functioning. The use of embedded dates in software is pervasive. There are five exposure areas related to Y2K or other date-related problem:

1. Operating system embedded dates
2. Application code embedded dates
3. Dates in the embedded software
4. Data (with potential command codes such as 9999 used as a program stop)
5. Dates embedded in interfacing commercial off-the-shelf software

The Y2K software audits had to examine all five potential fault areas above to assure that the software was free of the date fault. This type of problem can only be avoided by fully developing the program requirements; this means defining what the program "should not do" as well as what it "should do." The program requirements should also specify system desired behavior in the face of operational contingencies and potential anomalous program conditions. A good way to accomplish this is by performing failure modes and effects analysis on the high-level design (http://www.fmeainfocentre.com/). This will examine systematically the program responses to aberrant input conditions.

Good program logic will execute the desired function without introducing any logical ambiguities and uncertainties (e.g., different C compilers use a different order of precedence in mathematical operations). To assure proper compiling, it is best to use nested parentheses to assure the proper order of mathematical execution.

We now have an IT layer that envelopes our life and which we depend on. Internet use is pervasive and growing. Our offices, businesses, and personal lives depend on computer communications. We feel stranded when that service is not available, even for short periods. The most notorious outage was the

18-hour outage experienced by AOL users in August 1996 (http://www.cnn.com/TECH/9608/08/aol.resumes/). This resulted from an undetected operator error occurring during a software upgrade. The cause of failure was thought to be solely due to the upgrade. This problem confounding misled the analysts tracking the problem. Thus, 18 hours was required to isolate and correct the problem.

More important, many of our common services depend on computer controllers and communications. This includes power distribution, telephone, water supply, sewage disposal, and financial transactions as well as Web communications. Shocks to these services disrupt our way of life in the global economy. We are dependent on these services being up and functioning. The most critical and intertwining IT services that envelop our lives, either directly or indirectly, are:

1. Nuclear power
2. Medical equipment and devices
3. Traffic control (air, train, drive-by-wire automobiles, traffic control lights)
4. Environmental impact areas (smokestack filtration)
5. Business enterprise (paperless systems)
6. Financial systems
7. Common services (water, sewer, communications)
8. Missile and weapons systems
9. Remote systems (space, undersea, polar cap, mountaintop)
10. Transportation (autos, planes, subways, elevators, trains)

The *IEEE Spectrum* published a list of 30 software failures they called "the Hall of Shame" (http://www.spectrum.ieee.org/sep05/1685/failt1). Neumann [9] lists many key software failures.

SOFTWARE RELIABILITY: DEFINITIONS AND BASIC CONCEPTS

Failures can be defined as terminations of the ability of a functional unit to perform its required function. The user sees failures as a deviation of performance from the customer's requirements, expectations, and needs. A word-processing system that locks up and doesn't respond is a failure in the eyes of the customer. This is a failure of commission. A second type of failure occurs when the software is incapable of performing a needed function (failure of omission). This usually results from a "breakdown" in defining software requirements, which often is the dominant source of software and system problems. The customer is the one who determines what constitutes a failure. A failure signals the presence of an underlying fault that causes the failure to occur. A point should be made here: A fault are a logical construct (omission or commission) that leads to a software failure. Faults can often be corrected or circumvented in different ways. For example, hardware faults are corrected or averted by making software changes on a fielded system. System software changes are made since they can be done faster and cheaper than changing the hardware.

Faults are accidental conditions that make code susceptible to failure during operation. These are also called *latent defects* (these defects are hidden defects, whereas *patent defects* can be observed directly). The first time the newly developed F-16 fighter plane flew south of the equator, the navigational system of the plane caused it to flip over. This part of the navigational software had never been exercised prior to this trial. As long as the plane stayed in the northern hemisphere, there was no problem. A fault is merely a susceptibility to failure, it may never be triggered and thus a failure may never occur.

Fault triggers are those program conditions or inputs that activate a fault and trigger a failure. These failures typically result from untested branch, exception, or system conditions. Such triggers often result from unexpected inputs made to the system by the user, operator, or maintainer. For example, the user might inadvertently direct the system to do two opposing things at once, such as to go up and go down simultaneously. These aberrant or off-nominal input conditions stress the code robustness. The *exception-handling* aspects of the code are more complex than the operational code since there are a greater number of possible operational scenarios. Properly managing exception conditions are critical to achieving reliable software.

Failures can also arise from a program navigating untested operational paths or sequences that prove stressful for the code (e.g., some paths are susceptible to program timing failures). In these instances the desired sequence of program input conditions may violate the operational premise of the software and lead to erroneous output conditions. Timing failures are typically sensitive to the executable path taken in the program. There are so many sequences of paths through a program that it is typically said to be *combinatory-explosive analysis*. For example, 10 paths taken in random sequence lead to 2^{10} or 1024 combinations of possible paths. Practical testing considerations limit the number of path combinations that can be tested. There will always be some failure risk posed by the untested path combinations.

Errors are inadvertent omissions or commissions of the programmer that allows the software to misbehave relative to the users' needs and requirements. The relationship among errors, faults, and failures is shown in Figure 1. Ideally, a good software design should handle all inputs correctly under all environmental conditions. It should also correctly handle any errant or off-nominal input conditions. An example of this can be demonstrated in the following

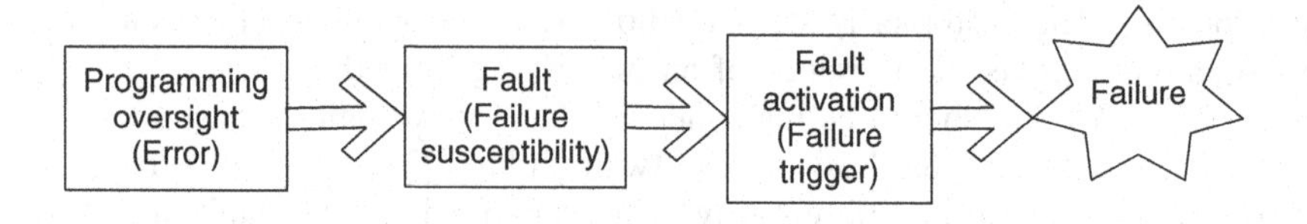

E.g., $F(x) = 1/(x + 234)$; well behaved except at $x = -234$
Programming error can occur anywhere in the process from requirements development to test

Figure 1 The path to failure.

simple arithmetic expression:

$$A + B = C \tag{1}$$

This operation is deterministic when inputs A and B are real numbers. However, what program response is desired when one of the inputs is unbounded, missing, imaginary, or textural? That is, how should the program handle erroneous inputs? Equation (1) would function best if the program recognized the aberrant condition. Then the software would not attempt to execute the program statement shown above but would report the errant input condition back to the user. Then this code would be handling the off-nominal condition in a robust manner. It would behave in the manner that the user would desire.

The mainline code usually does its job very well. Most often, the mainline code is sufficiently refined by the extensive testing that it has undergone. Software failures or code breakdowns occur when the software exception code does not properly handle a rarely experienced and untested condition. More frequently, the failure driver or trigger is an abnormal input or environmental condition. An example of a rarely occurring or unanticipated condition was the potential problems posed for programs failing to handle gracefully the millennium clock rollover. A classic example of improperly handling an off-nominal condition was the infamous "Malfunction 54" of a Therac 25 radiation therapy machine [9]. It injured some patients because the software controls did not satisfactorily interlock the high-energy operating system mode following a system malfunction. The operator had a workaround that allowed the radiation therapy machine to function, but unknown to the operator, the workaround defeated the software interlock. This allowed patients to be exposed to potentially lethal radiation levels.

There is a software development discipline that addresses directly the proper handling of branching and exception conditions. That is the "clean room" software design (http://www.dacs.dtic.mil/databases/url/key.hts?keycode=64). It assures validating the branches and exception conditions directly by routine checking of the exception conditions before it performs the mainline operational function. It does these checks at each operational step throughout the program.

Murphy and Gent [7] report that the biggest cause of problems for modern software stems from either requirements' deficiencies or improperly understood system interfaces. These are communications-based problems, which they label *system management failures*. They add a third system management driver: managing system changes. An example of gross miscommunications occurred when the Martian probe failed in the fall of 1999 (http://mars.jpl.nasa.gov/msp98/orbiter/). This failure was due to at least a couple of system management problems. First, there was a reported breakdown in communications between the propulsion engineers and the navigation scientists at a critical point in the design cycle. They were working under different assumptions that compromised design reliability. Second, the newspaper headlines publicized another program communication problem. The National Aeronautics and Space Administration (NASA) and its contractor had unconsciously been mixing metric and English dimensional units. The result helped misguide the $125 million space probe. These interface problems, often referred to as *interoperability problems*, are a pervasive threat to system reliability.

System management problems are truly communication problems. I have been a lifelong student of system failures (and successes). Today, for the most part, hardware is reliable and capable. Almost all significant system problems are traceable to a communications breakdown or too limited a design perspective on the programmer's or designer's part. Also, it needs to be recognized that programmers typically lack the "domain knowledge" of the application area for which they are designing the controlling software (e.g., the programmers know the C language or Java but are challenged when it comes to understanding the engineering aspects of a rocket control system). The system development people need to recognize these limitations and be proactive in supporting program requirements development. The programmers need to embrace the customer's "domain knowledge" in their system design. Davy and Cope [2] state: "In 2006 C. J. Davis, Fuller, Tremblay, & Berndt [1] found 'accurately capturing system requirements is the major factor in the failure of 90% of large software projects,' echoing earlier work by Lindquist [5], who concluded 'poor requirements management can be attributed to 71 percent of software projects that fail; greater than bad technology, missed deadlines, and change management issues.'"

Failure modes and effects analysis (FMEA) of the new software can also help to refine requirements. This analysis will help focus what the software is expected or desired to do as well as what behaviors are not desired. The software developer is first asked to explain the software's designed functionality. Then the person performing the FMEA will pose a set of abnormal input conditions that the software should handle. The software designer has been focusing on the positive path, and now the FMEA rounds out the remainder of the conditions that the software must accommodate. Samuel Keene has found that design FMEA analysis nominally takes two hours working with the software developer. This analysis has always led to the developer making changes to make the design more robust. Performing FMEA in conjunction with the developer is the key so that the developer will understand the problems and can resolve them quickly.

Designers must consider the possibility of off-nominal conditions that their codes must handle successfully. This is a proactive or defensive design approach. For example, in intensive numerical algorithms, there is always the possibility of divide-by-zero situations where the formula would go unbounded. When many calculations are involved or when one is dealing with inverting matrices, divide-by-zero happens too frequently. The code can defend itself by checking the denominator to ensure that it is nonzero before dividing. There is a performance cost to doing this operation. It is best when the designer is aware of the exposures and the trade-offs. A good design must be able to handle properly off-nominal inputs such as input absent; input late; input in wrong form, such as alphabetical when numerical is expected; and unbounded input. The catalog of anomalies can be built up with the individual contractor's experience.

Code changes are best made at scheduled intervals to assure a proper validation process of review, inspection, integration, and testing to assure that the new code fits seamlessly into the architecture. This practice precludes "rushed" changes being forced into the code. The new code changes will be regression-tested through the same waterfall development process (e.g., requirements validation, high-level code inspection, low-level code inspection, unit test, software integration and test, system test) as the base code. This change process preserves the original code development quality and maintains traceability to the design requirements. Using prototype and incremental releases helps reveal and refine program requirements. The spiral development model promotes progressive software validation and requirements refinement.http://en.wikipedia.org/wiki/Spiral_model

SOFTWARE RELIABILITY DESIGN CONSIDERATIONS

Software reliability is determined by the quality of the development process that produced it. Reliable code is exemplified by its greater understandability. Software understandability affects the software maintainability (and thus its reliability). The architecture will be more maintainable with reliable, fault-free code requiring minimal revision. The software developer is continually refining (maintaining) the code during development, and then the field support programmer maintains the code in the field.

Code reliability will be augmented by:

1. Increased cohesion of the code modules
2. Lowered coupling between modules, minimizing their interaction
3. Self-describing longer-variable mnemonic names
4. Uniform conventions, structures, naming conventions, and data descriptions
5. Code modularity optimized for understandability while minimizing overall complexity
6. Software documentation formatted for greatest clarity and understanding
7. Code commenting added to explain any extraordinary programming structures
8. Modular system architecture provides configuration flexibility and future growth
9. Restrict a single line of code to contain only a single function or operation
10. No use of negative logic, especially no double-negative logic in the design
11. Store any configuration parameters expected to change in a database to minimize the impacts of any changes on the design
12. Assure consistency of the design, source code, notation, terminology, and documentation
13. Maintain the software documentation (including flowcharts) to reflect the current software design level
14. Harmonize the code after each modification to assure that all the rules above are maintained

There are aging conditions in software that make it prone to failure. This occurs as its operating environment entropy increases. Consider the case of a distributed architecture air defense system, which had recognizance satellite receivers distributed across the country. These receivers experienced random failures in time. On one occasion the entire countrywide system was powered down and reset at one time. This system synchronization revealed an amazing thing. When the satellite receivers failed, they all failed simultaneously. Synchronizing the computer applications revealed the underlying common fault. A data buffer was tracking some particular rare transactions. It processed them and then passed them along. This register maintained a stack of all its input data, never clearing out the incoming data. So its overflow was inevitable. This overflow was a rare event, but the fault was revealed as soon as the system was synchronized.

Software reliability is strongly affected by the computer system environment (e.g., the length of buffer queues, memory leaks, resource contention). There are two lessons learned from software-aging considerations. First, software testing is most effective (likely to reveal bugs) when the distributed software is synchronously aged (i.e., restarted at the same time). At the other extreme, the assets distributed in the operational system will be more reliable if they are restarted asynchronously. Then processors are proactively restarted at advantageous intervals to reset the operating environment and restore the system entropy. This is called *software rejuvenation* or averting system state accretion failures (http://www.software-rejuvenation.com/). The system assets will naturally age asynchronously as the software assets are restarted following random hardware maintenance activities. Software running asynchronously helps preserve the architectural redundancy of system structure where one system asset can back up others. No redundancy exists if the backup asset succumbs concurrently to the same failure as the primary asset. Diversity of the backup asset helps preclude common-mode failures.

Software and system reliability will be improved by attention to the following factors:

1. Focus strongly and systematically on requirements development, validation, and traceability, with particular emphasis on system management aspects. Full requirements development also requires specifying things that the system should not do as well as desired actions (e.g., heat-seeking missiles should not boomerang and return to the installation that fired them). Quality functional development (QFD) is one tool that helps assure requirement completeness as well as fully documenting the requirements development process (http://www.shef.ac.uk/~ibberson/QFD-Introduction.html).

2. Institutionalize a "lessons learned" database and use it to avoid past problems and mitigate potential failures during the design process. Think defensively. Examine how the code handles off-nominal program inputs. Design to mitigate these conditions.

3. Prototype software releases are most helpful in clarifying the software's requirements. The user can see what the software will do and what it will not do. This operational prototype helps to clarify the user's needs and the developer's understanding of the user's requirements. Prototypes help the user and the developer gather experience and promote better operational and functional definition of the code. Prototypes also help clarify the environmental and system exception conditions that the code must handle. To paraphrase an old saw, "a prototype (picture) is worth a thousand words of a requirements statement." The spiral model advocates a series of prototypes to refine the design requirements recursively (http://www.sei.cmu.edu/cbs/spiral2000/february2000/Boehm/).

4. Build in diagnostic capability. Systems vulnerability is an evidence of omissions in our designs and implementation. Often, we fail to include error detection and correction (EDAC) capability as high-level requirements. The U.S. electrical power system is managed over a grid system that is able to use its atomic clocks for EDAC. These clocks monitor time to better than a billionth of a second. This timing capability's primary purpose is for proper cost accounting of power consumption. This timing capability provides a significant EDAC tool to disaggregate failure events that used to appear simultaneous. These precise clocks can also time reflections in long transmission lines to locate the break in the line to within 100 meters. The time stamps can also reveal which switch was the first to fail, helping to isolate the basic cause of the failure.

5. Carry out a potential failure modes and effects analysis to harden the system against abnormal conditions.

6. Failures should always be analyzed down to their underlying cause for repair and to prevent reoccurrence. To be the most proactive, the system software should be parsed to see if other instances exist where this same type of failure could result. The space shuttle did an excellent job of leveraging their failures to remove software faults. In one instance, during testing, they found a problem when system operations were suspended and the system failed upon restart. Their restart operation, in this instance, somehow violated the original design conception. The system could not handle this restart mode variation and it "crashed." This failure revealed an underlying fault structure. The space shuttle program parsed all 450 KSLOC (Thousands of source lines of code) of operational onboard code. Then two similar fault instances were identified and removed. So in this instance, a single failure led to the removal of three design faults. This proactive fault avoidance and removal procedure, called the *defect prevention process*, was the cornerstone of the highly reliable space shuttle program (http://www.engin.umd.umich.edu/CIS/course.des/cis565/lectures/sep15.html).

7. Every common-mode failure needs to be treated as critical, even if its system consequence is somehow not critical. Every thread that breaks the independence of redundant assets needs to be resolved to its root cause and remedied.

8. Study and profile the most significant failures. Understand the constructs that allowed the failure to happen. Y2K (the year 2000) is an example of a date defect. One should then ask if other potential date rollover problems exist. The answer is yes. For example, the global positioning system clock rolled over on August 22, 1999. On February 29, 2000 the leap year exception had to be dealt with. Here one failure can result in identifying and removing multiple faults of the same type.

9. Perform fault injection into systems as part of system development to speed the maturity of the software diagnostic and fault-handling capability.

OPERATIONAL RELIABILITY REQUIRES EFFECTIVE CHANGE MANAGEMENT

Beware: Small changes can have grave consequence. All too often, small changes are not treated seriously enough. Consequently, there is significant error proneness in making small code changes. Defect rate [11]:

1 line	50%
5 lines	75%
20 lines	35%

The defects here are any code change that results in anomalous code behavior (i.e., changes causing the code to fail). Often, small changes are not given enough respect. They are not analyzed or tested sufficiently. For example, DSC Communications Corp (Plano, Texas) signaling systems were at the heart of an unusual cluster of phone outages over a two-month period. These disruptions followed a minor software modification. The *Wall Street Journal* reported: "Three tiny bits of information in a huge program that ran several million lines were set incorrectly, omitting algorithms—computation procedures—that would have stopped the communication system from becoming congested, with messages.... Engineers decided that because the change was minor, the massive program would not have to undergo the rigorous 13-week (regression) test that most software is put through before it is shipped to the customer." Mistake!

Postrelease, program changes increase software's failure likelihood. Such changes can increase program complexity and degrade its architectural consistency. Because credit card companies have most of their volume between November 15 and January 15 each year, they have placed restrictions on telephony changes during that period, to limit interruptions to their business. This is good defensive planning.

EXECUTION-TIME SOFTWARE RELIABILITY MODELS

After the source code for a software system has been written, it undergoes testing to identify and remove defects before it is released to the customer. During this time, the failure history of the system can be used to estimate and forecast its reliability. The system's failure history is input to one or more statistical models, which produce as their output quantities related to the software's reliability. The failure history takes one of two forms:

- Time between subsequent failures
- Number of failures observed during each test interval, and the length of that interval

The models return a probability density function for either the time to the next failure or the number of failures in the next test interval. This is then used to estimate and forecast the reliability of the software being tested. Further details on these and additional models may be found in refs. [3, 4, 6, 8]. The most popular software reliability estimation tool, the CASRE (computer-aided software reliability estimation) tool suite, developed with funding from the U.S. Air Force Operational Test and Evaluation Center, is a stand-alone Windows-based tool that implements 10 of the more widely used execution time software reliability models. CASRE's goal was to make it easier for nonspecialists in software reliability to apply models to a set of failure history data and make predictions of what the software's reliability will be in the future. Perhaps the greatest limitation of execution time software reliability models is that they cannot be applied until the testing phases of a development effort. Therefore, these execution-time models are limited in promoting early program reliability decisions and trade-offs between development methods, budget, and reliability. A new reliability tool that is based on the quality of the development process is discussed next.

SOFTWARE RELIABILITY PREDICTION TOOLS PRIOR TO TESTING

Software failure rates are a function of the development process used. The more comprehensive and better the process is, the lower the fault density of the resulting code. There is an intuitive correlation between the development process used and the quality and reliability of the resulting code, as shown in Figure 2. The software development process is largely an assurance and bug removal process. Eighty percent of the development effort is spent removing bugs. The greater the process and assurance emphasis, the better the quality of the resulting code. Several operational field data points have been found to support this relationship. The operational process capability is measured by several techniques. The best known technique is the capability maturity model used by the Software Engineering Institute (SEI). It grades development processes from a 1 to 5 (or I to V) rating. The higher rating indicates better processes. Thus, this process measure will be used to project the latent fault content of the code developed.

Figure 2 shows that the defect density rate improves (decreases) as the development team's capability improves. The nominal fault densities range from six faults per KSLOC for a level 5 capability process to 0.1 fault for a level 5 capability process. Also, the higher CMM-level development organizations will have a more consistent process, resulting in a tighter distribution of the fault density and failure rate observed for the fielded code. They will have fewer outliers and greater predictability of the latent fault rate.

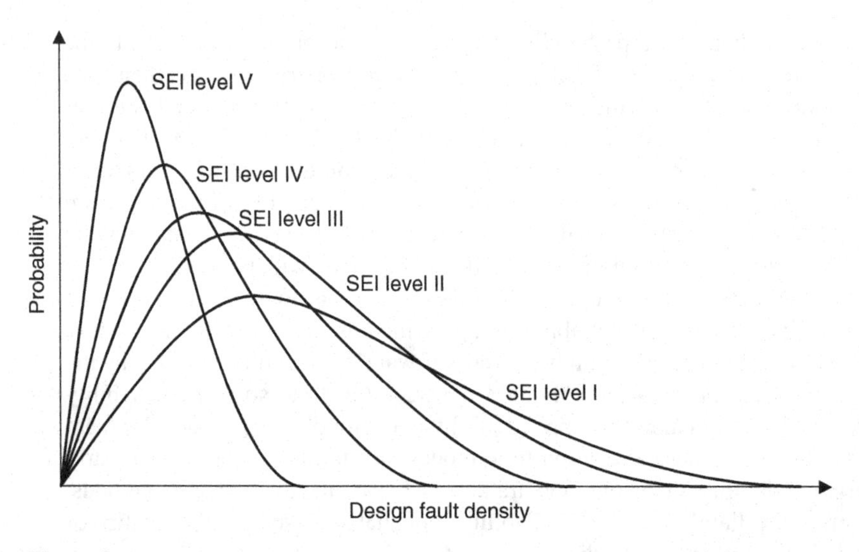

Figure 2 Keene development process fault model as a function of the underlying development process capability.

Keene's development process prediction model correlates the latent fault content delivered with the development process capability [3, App. F; 10]. This model can be used in the program planning stages to predict the operational software reliability. The model requires user inputs of the following parameters:

1. Estimated KSLOCs of deliverable code.
2. SEI capability level of the development organization.
3. SEI capability level of the maintenance organization.
4. Estimated number of months to reach maturity after release (historical).
5. Use hours per week of the code.
6. fault activation (estimated parameter). This represents the average percentage of seats of system users that are likely to experience a particular fault. This is especially important (much less than 100%) for widely deployed systems such as the operating system AIX, which has over 1 million seats. This ratio appears to be a decreasing function over time that the software is in the field. Faults discovered early tend to infect a larger ratio of total systems. Faults removed later are more elusive and specialized to smaller domains (i.e., have a smaller, and stabilizing, fault activation ratio). A fault activation level of 100% applies when there is only one instance of the system.

Fault latency is the number of times a failure is expected to reoccur before being removed from the system. It is a function of the time it takes to isolate a failure, design and test a fix, and field the fix that precludes its reoccurrence. This factor is based on the process capability level and development history of similar products.

The Keene development process model provides a ready method of estimating fault content and the resulting failure rate distribution at requirements planning time. It rewards a developing organization that demonstrates better process capability with lower fault content and projected better field failure rates. This model requires developers to know some things about their experience with released code, to fill in all the model parameters. It is now being used by several defense and aerospace contractors.

REFERENCES

[1] Davis, C. J., Fuller, R. M., Tremblay, M. C., and Berndt, D. J., Communication challenges in requirements elicitation and the use of the repertory grid technique, *J. Comput. Inform. Syst*., vol. 78, 2006.

[2] Davy, B., and Cope, C., Requirements elicitation—What's missing? *Issues Inform. Sci. Inform. Technol*., vol. 5, 2008, p. 543.

[3] *Recommended Practice on Software Reliability*, IEEE 1633–2008, IEEE, Piscataway, NJ, June 2008.

[4] Kan, S. H., *Metrics and Models in Software Quality Engineering*, Addison-Wesley, Reading, MA, 1995, p. 192 (Rayleigh model discussion).

[5] Lindquist, C., Required: Fixing the requirements mess: The requirements process, literally, deciding what should be included in software, is destroying projects in ways that aren't evident until it's too late. Some CIOs are stepping in to rewrite the rules, *CIO*, vol. 19, no. 4, 2005, p. 1

[6] Lyu, M. R., Ed., *Handbook of Software Reliability Engineering*, Computer Society Press, Los Alamitos, CA, and McGraw-Hill, New York, 1995.

[7] Murphy, B., and Gent, T., Measuring system and software reliability using an automated data collection process, *Qual. Reliab. Eng. Int*., CCC 0748-8017/95/050341-13pp., 1995.

[8] Musa, J. D., Iannino, A., and Okumoto, K., *Software Reliability: Measurement, Prediction, Application*, McGraw-Hill, New York, 1987.

[9] Neumann, P. G., *Computer Related Risks*, ACM Press, New York, 1995.

[10] Peterson, J., Yin, M.-L., and Keene, S., Managing reliability development and growth in a widely distributed, safety-critical system, Practical Papers, 12th Annual ISSRE, Hong Kong, China, Nov. 27–30, 2001.

[11] Edwards, William, "Lessons Learned from 2 Years Inspection Data", Crosstalk Magazine, No. 39, Dec 1992, cite: Weinberg. G. "Kill That Code!", IEEE Tutorial on Software Restructuring, 1986, p. 131.

Chapter 4

Reliability Models

Louis J. Gullo

INTRODUCTION

Models are developed to simulate actual behavior or performance and to reflect or predict reality. A modeling methodology is needed to fully understand what models are and how they are useful. Maier and Rechtin [1, p. 11] state: "Modeling is the creation of abstractions or representations of the system to predict and analyze performance, costs, schedules, and risks, and to provide guidelines for systems research, development, design, manufacture, and management. Modeling is the centerpiece of systems architecting—a mechanism of communication to clients and builders, of design management with engineers and designers, of maintaining system integrity with project management, and of learning for the architect, personally."

The goal of a reliability model is to set expectations on performance and reliability, and to represent reality accurately as closely as possible for the purpose of predicting or assessing future performance and reliability. A key focus in developing models is to ensure that no mission-aborting (mission-critical) component will fail during a required mission time. This focus could be extrapolated into no mission failures during the expected lifetime (over a number of missions). This goal is accomplished by identifying design weaknesses using experimental design modeling, analysis, and testing, to improve design for reliability such that mission interruptions are minimized or reduced over the anticipated life, and to reduce unscheduled downtime and increase system availability.

There are many times during a system or product life cycle when reliability modeling has value. First, in the system or product concept and development

Design for Reliability, First Edition. Edited by Dev Raheja, Louis J. Gullo.

phase, modeling is valuable to help system architects and product designers to decide where and how much redundancy and fault tolerance is needed to achieve life-cycle requirements. Modeling may also be employed after initial development when a major redesign is being considered for product feature enhancements and technology refresh. If organizations want to continue using the systems beyond the expected life, they should conduct risk analysis with the aid of reliability models to compare the reliability of the design with the maintenance cost to refurbish. Reliability modeling helps organizations make the decision to determine when it is not economically feasible to support a system or refurbish a system beyond its expected lifetime.

One type of reliability model is the reliability prediction model. Reliability prediction models may be physics-of-failure (PoF) models, failure rate models, or mean-time-between-failures (MTBF) models. Reliability prediction models may take the form of probability density functions or cumulative density functions. Reliability prediction models may be useful for comparing the reliability of two different designs. A common approach for comparing the reliability of different design options is described in MIL-HDBK-217 [2], the military handbook for the reliability prediction of electronic equipment. The purpose of MIL-HDBK-217 is to establish and maintain consistent and uniform methods for estimating the inherent reliability (i.e., the reliability of a mature design) of military electronic equipment and systems. It provides a common basis for reliability predictions during acquisition programs for military electronic systems and equipment. The handbook is intended to be used as a tool to increase the reliability of the equipment being designed. It also establishes a common basis for comparing and evaluating reliability predictions of related or competitive designs. The military handbook includes component failure rate models for various electronic part types. A typical part stress failure rate model is shown in this example. The part stress failure rate, λ_p, for fixed resistors is

$$\lambda_p = \lambda_b(\pi_E \times \pi_R \times \pi_Q) \tag{1}$$

where

π_b = base failure rate (a function of temperature and stress)
π_E = environmental factor
π_R = resistance factor
π_Q = quality factor

Reliability models may be created to predict reliability from design parameters, as is the case when using MIL-HDBK-217, or they may be created to predict reliability from experimental data, test data, or field performance failure data. One type of reliability model, which is developed based on design parameters, is called a *defect density model*. These types of reliability models have been used extensively for software reliability analysis. Various types of software reliability models are provided in IEEE 1633–2008 [3], a practice recommended for software reliability engineering. Software reliability models are based on design parameters, the maturity of a design organization's processes, and software code characteristics, such as lines of code, nesting of loops, external references, flow in and flow out, and inputs and outputs, to estimate the number of defects in software. Reliability models statistically correlate defect detection data with known functions or distributions, such as an exponential probability distribution function. If the data correlate with the known distributions, these distributions or functions can be used to predict future behavior. This does not mean that the data correlate exactly with the known functions but, rather, that there are patterns that will surface. The similarity modeling method, discussed later in the chapter, estimates the reliability of a new product design based on the reliability of predecessor product designs, with an assessment of the characteristic differences between the new design and the predecessor designs. This is possible only when the previous product designs are sufficiently similar to the new design and there is sufficient high-quality data. The use of actual performance data from test or field operation enables reliability predictions to be refined and "measurements" to be taken. The higher the frequency and duration of the correlation patterns, the better the correlation. Repeatability of patterns that exist will determine a high level of confidence and accuracy in the model. Even with a high level of confidence and good data correlation, there will be errors in models, yet the models are still useful.

George Box is credited with the saying: "All models are wrong, some are useful." Box is best known for the book *Statistics for Experimenters: Design, Innovation and Discovery* [4], which he coauthored with J. S. Hunter and W. G. Hunter. This book is a highly recommended resource in the field of experimental statistics and is considered essential for work in experimental design and analysis. Box also co-wrote the book *Time Series Analysis: Forecasting and Control* [5] with G. M. Jenkins and G. Reinsel. This book is a practical guide dealing with stochastic (statistical) models for time series and their use in several applications, such as forecasting (predicting), model specifications, transfer function modeling of dynamic relationships, and process controls, to name just a few applications. Box served as a chemist in the British Army Engineers during World War II, and following the war, received a degree in mathematics and statistics from University College London. As a chemist trying to develop defenses against chemical weapons in wartime England, Box needed to analyze data from experiments. Based on this initial work and his continuing studies and efforts thereafter, Box became one of the leading communicators of statistical theory.

As a co-developer of PRISM (a registered trademark of Alion Science and Technology) along with Bill Denson from the Reliability Analysis Center, Samuel Keene believed (in the 1990s) that the underlying thesis in this reliability model was that the majority of significant failures tie back into the process capability of the developing organization. He worked in the MIL-HDK-217 working group during that time. He completed his work with a view that parts are the underlying causes of system failures. Then to his surprise he found that 95% of the problems on production systems were traceable to a few parts experiencing a specific cause failure, with the majority of failures related to system-level causes not related to the inherent reliability of the parts.

Many types of reliability models are described in this chapter. Reliability models are developed for systems, hardware, and software to develop optimized solutions to reliability problems. System reliability models may be classified as repairable models or non-repairable models. An excellent reference for probabilistic modeling of repairable systems is by Ascher and Feingold [11]. An overview of models and methods for solving reliability optimization problems since 1970's is presented by Kuo and Prasad [12]. A common form of system reliability models is a Reliability Block Diagram (RBD).

RELIABILITY BLOCK DIAGRAM: SYSTEM MODELING

System reliability block diagrams (RBDs) and models start with reliability allocations derived and decomposed from requirements. These allocations are apportioned to the blocks of the model. The model is updated with reliability predictions based on engineering analysis of the design to validate the model initially. The predictions are updated with actual test and field reliability data to validate the model further.

What is an RBD?

- An RBD is a logical flow of inputs and outputs required for a particular function.
- There are three types of RDB diagrams:
    - Series RDB
    - Parallel RBD (includes active redundancy, standby, or passive redundancy)
    - Series–parallel RBD

Examples of each RDB type are shown later in the chapter.

Two of the most frequently used RDB models are basic reliability models and mission reliability models. These two types of RBM models are discussed in detail in MIL-STD-756B [6] and by O'Connor [7]. The basic reliability model may be in the form of a series RBD model. A basic reliability model is a static model that demonstrates how to determine the cost of spares provisioning to support the maintenance concept throughout the life cycle. The basic reliability model is a serial model that accumulates the failure rates of all components of a system. This type of model considers the inherent failure rate of each hardware or software component or module. It helps the analyst recognize the fact that adding more hardware or software to a system decreases the reliability of the system and increases the cost to maintain the system.

The mission reliability model is another static model that demonstrates the system reliability advantages when using a fault-tolerant design that includes redundancy and hot sparing concepts to meet mission reliability requirements. A mission reliability model may be in the form of a parallel or a series–parallel RBD. The mission reliability requirement may be stated in terms of the probability of a mission-critical failure during the mission time. Mission time may be the time it takes to drive from point A to point B, or the time it takes for a missile to reach its target. The mission reliability model could be a combined serial–parallel RBD model which recognizes that adding more hardware or software does not necessarily decrease the reliability of the system, but increases the cost to maintain the system. In many cases for a parallel RBD model, adding redundant hardware will increase the reliability of the system and increase the maintenance costs during normal operation. The increase in maintenance cost results from the additional hardware to remove and replace in the system when a system failure occurs, from more spare hardware to store and replenish, and from higher logistics times to transport and supply the spares and perform the maintenance tasks.

EXAMPLE OF SYSTEM RELIABILITY MODELS USING RBDs

- The reliability of a complex system depends on the reliability of its parts.
- A mathematical relationship exists between the parts' reliability and the system's reliability.
- If a system consists of n units, with reliability R_i for unit i, then in simple terms, the system reliability, R_{sys}, is given by

$$\mathrm{R_{sys}} = \begin{cases} R_1 R_2 \cdots R_n & \text{when the units are in series} \\ 1 - F_1 F_2 \cdots F_n & \text{when the units are in parallel,} \\ & \quad \text{where } F_i = 1 - R_i \end{cases}$$

Case 1

A series RBD model (Figure 1) consists of four components, A to D, each with a reliability of 0.99 for the conditions stated and time t. The four components

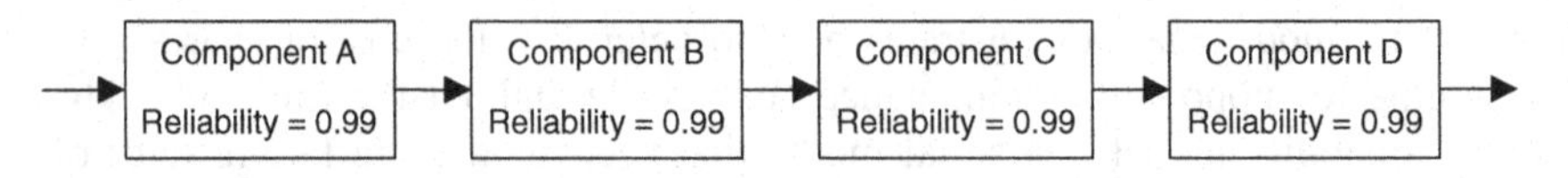

Figure 1 Example of a series RBD model.

are connected in series and an equation is written to show the mathematical expression for this model:

$$R_{\mathrm{sys}} = \prod R_i = (0.99)^4 = 0.96 \tag{2}$$

In this case, a series reliability model results in a total system reliability that is less than the reliability of a single component.

Case 2

A parallel RBD model (Figure 2) consists of three components, A to C, that are connected in active parallel with each component's $R_i = 0.99$. An equation also shows the mathematical expression for this model:

$$\begin{aligned} R_{\mathrm{sys}} &= 1 - \left(\prod F_i\right) = 1 - (0.01)^3 = 1 - 0.000001 \\ &= 0.999999 \end{aligned} \tag{3}$$

where $F_i = 1 - R_i$ and $R_i = 0.99$. In this case, a parallel reliability model results in a total system reliability that is greater than the reliability of each of the components.

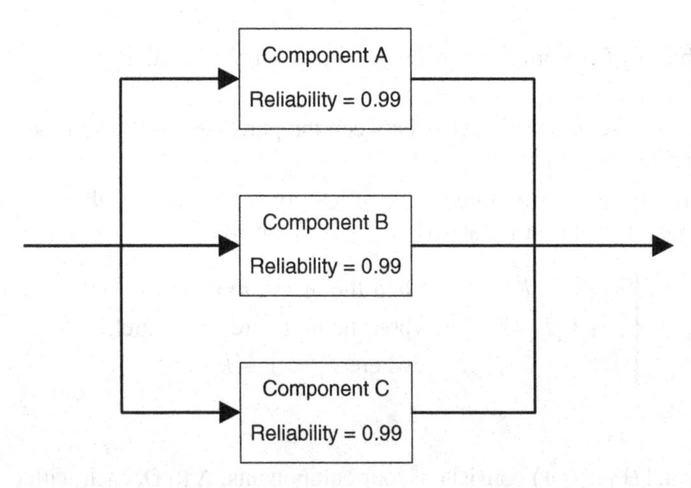

Figure 2 Example of a parallel RBD model.

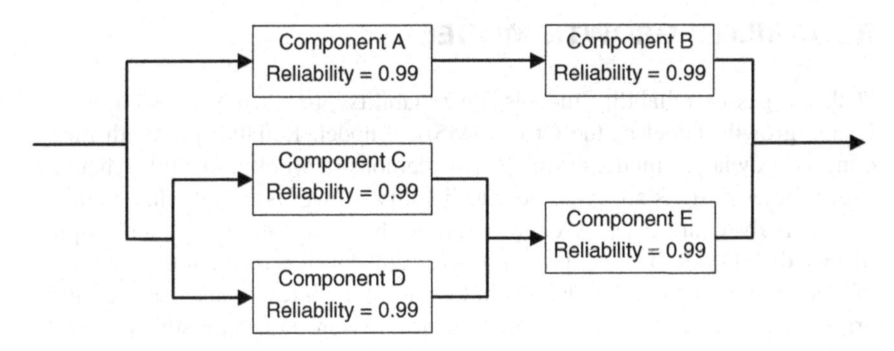

Figure 3 Example of a series-parallel RBD model.

Case 3

A series–parallel RBD model (Figure 3) consists of five components, A to E, that are connected in an active series–parallel network with each component's $R_i = 0.99$. An equation shows the mathematical expression for this model:

$$\begin{aligned} R_{\text{sys}} &= 1 - (1 - R_1 R_2)\{1 - [1 - (1 - R_3)(1 - R_4)]R_5\} \\ &= 1 - 0.000201 = 0.999799 \end{aligned} \tag{4}$$

where $F_i = 1 - R_i$ and $R_i = 0.99$. In this case, a series–parallel reliability model results in a total system reliability that is greater than the reliability of any one component.

System Combined

- If the specified operating time, t, is made up of different time intervals (operating conditions), t_a, t_b, t_c, etc.:

$$R(t_a) = e^{-\lambda_a t_a}$$
$$R(t_b) = e^{-\lambda_b t_b}$$
$$R(t_c) = e^{-\lambda_c t_c}$$

- The system combined reliability is shown in equation (5) and is simplified in equation (6):

$$R(t) = R(t_a)R(t_b)R(t_c) = e^{-\lambda_a t_a} e^{-\lambda_b t_b} e^{-\lambda_c t_c} \tag{5}$$
$$R(t) = e^{-(\lambda_a t_a + \lambda_b t_b + \lambda_c t_c)} \tag{6}$$

Note: In this example, where the reliability distribution is an exponential distribution, the assumption is made that the failure rate is constant over time.

RELIABILITY GROWTH MODEL

"Other types of reliability models are reliability growth models such as the Duane growth model or the Crow AMSAA model. Reliability growth modeling is a dynamic modeling method to demonstrate how reliability changes over time as designs are improved. Reliability growth is useful when there is a Test Analyze and Fix (TAAF) program to discover design flaws and implement fixes [11]. Reliability growth curves are developed to show the results of design improvements as demonstrated in a growth test, whenever the interarrival times tend to become larger for any reason. Other reliability growth models available in the literature include the Nonhomogeneous Poisson Process (NHPP) models, including the Power Law model, Cox-Lewis (Cozzolino) model, McWilliams model, Braun-Paine model, Singpurwalla model, Gompertz, Lloyd-Lipow, and logistic reliability growth models [8,11]. A reliability growth model is not an RBD model. Reliability growth models are designed to show continuous design reliability improvements. Reliability growth models assess the rate of change in reliability over time, which can be used to show additional improvements beyond a single RBD model."

SIMILARITY ANALYSIS AND CATEGORIES OF A PHYSICAL MODEL

Reliability modeling using similarity analysis is discussed in refs. [6, 9, 10]. The similarity analysis method should define physical model categories to compare new and predecessor end items or assemblies. These categories are shown in Table 1. The first five categories are part-type component-level categories that quantify the field failures due to component failure and may be partitioned to various levels of detail. The next two categories are design and manufacturing processes. Additional categories may be added for equipment-specific items not related to part type or process categories.

In this example, failure causes are divided into physical model categories. Categories 1 to 5 are for failures in service due to component failure; the next two categories are for processes (design and manufacturing) that can be controlled by the end item supplier, and additional categories (categories 8 and higher) can be added for product-specific items. Manufacturing-induced component failures are categorized under manufacturing processes (category 6), and misapplication and overstress are categorized under the design processes (category 7).

Equation (7) is used to calculate the total product failure rate and the projected mean time between failures (MTBF). The total product failure rate is computed by summing the failure rates for categories 1 through 8. Projected

Table 1 Example of Physical Model Categories

Category 1	Low-complexity passive parts (resistors, capacitors, and inductors)
Category 2	High-complexity passive parts (transformers, crystal oscillators, and passive filters)
Category 3	Interconnections (connectors, printed wiring boards, and solder joints)
Category 4	Low-complexity semiconductors (discretes, linear ICs, and digital ICs)
Category 5	High-complexity semiconductors (processors, memory, and field-programmable gate arrays and application-specific integrated circuits)
Category 6	Manufacturing process
Category 7	Design process
Category 8 and higher	Other failure causes which are specified by the user to describe failure mechanisms that do not fit categories 1 to 7, or that the analyst wishes to track separately due to high frequency of occurrence, or life-limited failures such as lamps or switches, or design modifications that correct design flaws

MTBF is computed by taking the inverse of the total product failure rate:

$$\text{total product failure rate}(\lambda) = \left[\sum_{C=1}^{5}\left(\sum_{L=1}^{N} Q_{L,C}\right)\lambda_C\right] + (F_M\lambda_6) + (F_D\lambda_7) + \left(\sum_{L=1}^{N} Q_{L,8}\right)\lambda_8 \quad (7)$$

where

$Q_{L,C}$ = part quantities for line replaceable unit (LRU), assembly L, and physical model category C

L = one LRU or field-replaceable unit (FRU)

C = one physical model part-type component-level category (categories 1 to 5)

N = total number of LRU, assembly, or functional levels in the assessment.

λ_C = expected category failure rate for physical model category C

F_M = process factor for the manufacturing process failures in category 6

F_D = process factor for the design process failures in category 7

λ_6 = expected category failure rate for the manufacturing process—physical model category 6

λ_7 = expected category failure rate for the design process—physical model category 7

λ_8 = expected category failure rate for physical model category 8

Formula representation (7) assumes one user-defined physical model category (category 8). If additional user-defined categories are required (e.g., category 9, 10, etc.), they will be treated in the same manner as category 8.

Companies must understand the inherent design reliability of complex systems where RDBs and similarity analysis methods are not adequate. To compute the reliability of complex systems, companies may use Monte Carlo simulations and models included with RDBs, and Markov chains and models to supplement RDBs.

MONTE CARLO MODELS

Monte Carlo models and simulations provide a means to assess future uncertainty for complex physical and mathematical systems. They are often applied to RBDs in the development of reliability models for complex systems. Monte Carlo analyses produce outcome ranges based on probability, incorporating future uncertainty. Monte Carlo methods combine analysis of actual historical data over a sufficient period of time (e.g., one year) with analysis of a large number of hypothetical histories (e.g., 5000 to 10,000 iterations) for a projected period of time into the future. These models rely on repeated random sampling to compute reliability. These methods are useful when it is infeasible or extremely difficult to compute an exact result with a deterministic algorithm. Monte Carlo methods are used in mathematics, such as for the evaluation of multidimensional integrals with complicated boundary conditions and in reliability analyses to correlate electrical and environmental stresses to time-to-failure data. Variation in failure rates could be analyzed based on multivariant stress conditions, such as high electrical stress with high-thermal-stress conditions, high electrical stress with low-thermal-stress conditions, and low electrical stress with low-thermal-stress conditions. These types of Monte Carlo analyses may also be known as four- or five-corner analyses, depending on the number of stress parameters that are evaluated simultaneously.

MARKOV MODELS

For systems subject to the dynamic behavior of functional dependence, sequence dependence, priorities of fault events, and cold spares, Markov models and Markov chain methods are used for reliability modeling. Markov chain methods consider each component failure, one by one. Markov models are very useful for understanding the behavior of simple systems, but as the system becomes more complex, there is a major drawback. As the number of states increase significantly, the number of probabilities to be calculated increases, so that the Markov method becomes very cumbersome compared to RDB models.

Consider a simple system with one component in service and one spare component in stock. There are three possible states:

- State 1: both components serviceable, one component in service and one component in supply
- State 2: one component serviceable, one component unserviceable and in repair
- State 3: both components unserviceable

For any given time interval, the system may remain in one state or change state once, twice, or more times, depending on the length of time chosen. If we consider a time interval where the system is initially in state 1, all the possibilities can be shown diagrammatically as in Figure 4.

Typical inputs and outputs from a Markov model are:

State = system state

λ = failure rate

μ = repair rate

P = probability

P_{ii} = probability of system remaining in state i

P_{ij} = probability of system changing state from state i to state j

Further explanation of Monte Carlo methods, Markov methods, and reliability prediction models may be found in IEEE 1413.1 [10].

Modeling a system is consistent with the design for reliability (DFR) knowledge thread of the book: that modeling a system that includes DFR eliminates the potential for mission-aborting failures. Electronic designs will have random isolated incidences that will not warrant design changes, such as failures within a six sigma defect envelope: for example, 4,000,000 devices tested and one integrated circuit (IC) component fails because of a broken bond wire that was tested previously. These defects may result in system-degraded performance but not mission-critical failures. Modeling a system

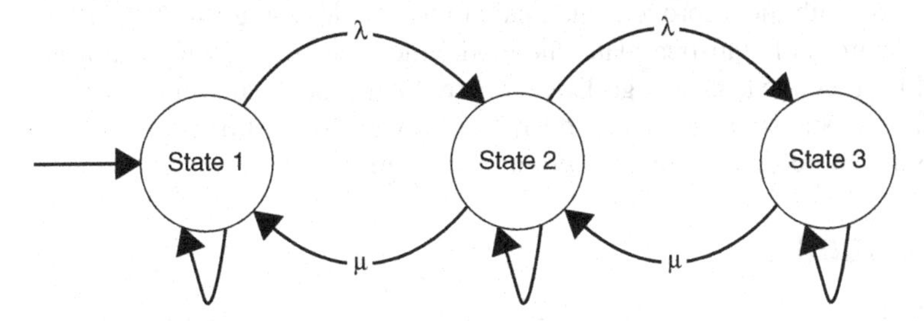

Figure 4 Example of a 3-state Markov model.

that allows for scheduled or preventive maintenance or condition-based maintenance is to develop views and viewpoints that incorporate an understaning of what components of the system warrant replacement and provide for an allowable failure rate while the system is performing its mission. More details are available in the references.

Reliability models which assume that a system failure model is merely the sum of its parts failure rates are inaccurate, yet these methods exist today. Electronic part failures account for 20 to 50% of total system failures. For nonelectronic parts, the failure distribution percentage is even smaller. The majority of the failure distribution is related to customer-use or customer-misuse conditions or defects in the system requirements. The reason that parts are usually blamed for many, if not all, failures is because it is easy to allocate failures to part numbers in showing proof of root-cause analysis and drilling down to determine the physics of failure. As we see it, the problem is not the part but the use of the part or out-of-spec conditions that caused the failure.

In a good failure reporting, analysis, and corrective action system (FRACAS), actual performance data from testing and field experience is useful to modify or update the reliability models and improve confidence in the models. Failure cause codes will be created for such failure causes as design failures, design process failures, customer-induced failures, manufacturing process and workmanship defects, supplier process defects, and supplier part design failures, as a minimum. These failure-cause categories are used to revise reliability models and build the model accuracy. All these failure causes should be correctly coded in the FRACAS database and correlated with their appropriate failure-cause categories which are applied in the models (discussed earlier). For example, a failure caused by the electrical assembly designer's misapplication of a part in a circuit causing electrical overstress should be coded as a design-related failure and not as a supplier part issue. Care must be taken during root-cause failure analysis to determine the exact failure mechanisms and attribute failure causes to the proper cause codes for future modeling and data analysis. For an effective useful model, all the causes of system failures should be allocated a failure rate. The early development models should be improved constantly using all available data collected from testing and field performance. With the improvements made to the model using FRACAS data and other sources of statistical data, the model increases in accuracy and usefulness. In the context of George Box's quote, "All models are wrong, some are useful," as statistical data are added to a model, its usefulness increases and becomes more correct, or less wrong and more useful.

REFERENCES

[1] Maier, M. W., and Rechtin, E., *The Art of Systems Architecting*, 2nd ed., CRC Press, Boca Raton, FL, 2002.

[2] *Military Handbook for the Reliability Prediction of Electronic Equipment*, MIL-HDBK-217, Revision F, Notice 2, U.S. Department of Defense, Washington, DC, Feb. 1995.

[3] *Recommended Practice on Software Reliability*, IEEE 1633–2008, IEEE, Piscataway, NJ, June 2008.

[4] Box, G., Hunter, J., and Hunter, W., *Statistics for Experimenters: Design, Innovation and Discovery*, 2nd ed., Wiley-Interscience, Hoboken, NJ, 2005.

[5] Box, G., Jenkins, G. M., and Reinsel, G., *Time Series Analysis: Forecasting and Control*, 4th ed., Wiley, Hoboken, NJ, 2008.

[6] *Military Standard on Reliability Modeling and Prediction*, MIL-STD-756B, U.S. Department of Defense, Washington, DC, 1981.

[7] O'Connor, P., *Practical Reliability Engineering*, 4th Ed., Wiley, Hoboken, NJ, 2002.

[8] Kececioglu, B. D., *Reliability Engineering Handbook*, Vols. 1 and 2, 2nd ed., Prentice Hall, Upper Saddle River, NJ, 1991.

[9] Johnson, B., and Gullo, L., Improvements in reliability assessment and prediction methodology, presented at the Annual Reliability and Maintainability Symposium, 2000.

[10] *IEEE Guide for Selecting and Using Reliability Predictions*, IEEE 1413.1-2002 (based on IEEE 1413), IEEE, Piscataway, NJ.

[11] Ascher, H., and Feingold, H., *Repairable Systems: Modeling, Inference, Misconceptions and Their Causes*, Marcel Dekker, New York, 1984.

[12] Kuo, W., and Prasad, V. R., An annotated overview of system-reliability optimization, *IEEE Trans. Reliab.*, vol. 49, no. 2, 2000, pp. 176–187.

Chapter 5

Design Failure Modes, Effects, and Criticality Analysis

Louis J. Gullo

INTRODUCTION TO FMEA AND FMECA

Failure modes and effects analysis (FMEA) is a complex engineering analysis methodology used to identify potential failure modes, failure causes, failure effects and problem areas affecting the system or product mission success, hardware and software reliability, maintainability, and safety. FMEA provides a structured process for assessing failure modes and mitigating the effects of those failure modes through corrective actions. When FMEA is performed on hardware in collaboration with the electrical circuit designer or the mechanical design engineer, it is very useful in effecting design improvements. The success of FMEA depends on collaboration between the FMEA analyst and the designers and stakeholders. This involves conducting open dialogue sessions, which are especially valuable in uncovering and resolving single-point failure modes at discrete levels of the system or product design that have an unacceptably high probability of occurrence or a critically severe failure effect that causes personnel injury or high system repair cost due to loss of system functionality. Elimination of single-point failures is one of the primary concerns of an analyst performing FMEA. For key findings, design changes should be incorporated to reduce the severity of the failure effects or to minimize

Design for Reliability, First Edition. Edited by Dev Raheja, Louis J. Gullo.

the probability of occurrence of the particular failure mode or to increase the detectability of the failure mode in order to reduce risk.

Failure modes, effects, and criticality analysis (FMECA) is an analysis of a system or product similar to FMEA, with the addition of quantitative criticality analysis using equations to calculate the criticality of each failure mode. The FMECA is very useful when applied to a design, for assessing the failure mode criticalities and comparing design failure criticalities with each other and ranking the failures, or determining the criticality relative to a benchmark criticality or threshold level. Criticality helps to rank priorities objectively for follow-up studies, maintenance task analysis, and design improvements. Criticality is useful for calculating the risk priority number, which is valuable for ranking the failures in terms of priorities for taking actions to mitigate the risks, such as conducting design changes.

DESIGN FMECA

One type of FMECA is design FMECA (D-FMECA). FMECA of a design is analysis of system or product performance considering what happens when or if a failure occurs. This type of FMECA is performed by examining assembly drawings, part datasheets, electrical schematics, and specifications. Design FMECA does not include analysis of manufacturing-related failures, workmanship defects, and random isolated incidences related to variations in assembly or component supplier–build processes.

D-FMECA functions as a living document during the development of product or system hardware and software design. The value of FMECA is determined by the early identification of all critical and catastrophic subsystem or system failure modes so that they can be eliminated or minimized through design improvements early in development, prior to production and delivery to the customer. It is important to update the data contained in a D-FMECA document continually with actual failure modes and effects data from testing and actual field applications, to keep pace with the evolving design so that it can be used effectively throughout the development and sustainment phases of the product or system life cycle. Furthermore, the results of D-FMECA are valuable for logistics support analysis reports and other tasks that may be part of an integrated logistics support plan.

The purpose of D-FMECA is to analyze a system or product design to determine the results or effects of failure modes on system or product operation, and to classify each potential system failure according to its severity, failure occurrence, detection method, and risk priority number (RPN), which is a value that summarizes the effects of the severity, occurrence, and detectibility. Several references provide approaches to calculate the failure criticality or RPN [0–9]. Each failure mode identified will be classified with an RPN, which will be used in the design process to assign priorities for design-corrective actions. Many times, an RPN limit is set to a predefined threshold, such as 6 on a scale of 0 to 9. RPNs are calculated for each failure mode and compared to the RPN limit, and a design change decision is made based on scoring of failure modes. For example, a failure mode with an RPN>6 must be eliminated or, as a minimum, its effects reduced. An alternative method is to rank all the listed entries by RPN and select the higher-value RPN entries for follow-up.

Most sources and references have previously defined RPN as a multiplicative calculation method. This method may result in large numbers within a wide range of values which are separated by several orders of magnitude. The casual analyst may deem this method to be too complicated for rapid analysis, to quickly differentiate failure modes that designers should be concerned about versus failure modes that designers may defer fixing until later in the development cycle. In this chapter the approach to calculating RPN is to provide an additive model, which is easier for mainstream us and simpler for the casual analyst to grasp.

RPNs assist the designer in prioritizing the failure modes to fix immediately. Risk mitigation techniques are developed to correct the high-risk single-point failures first: to offset the risks, to reduce the risks, or to eliminate the risks (also known as risk avoidance). The design improvements could be planned as scheduled system or product enhancements incorporated at a later date or, if the severity warrants it, incorporated immediately.

The goal of FMECA is zero design-related single-point failures. Where single-point design-related failures cannot be eliminated, the effects of single-point failures and the likelihood of occurrence of single-point failures should be minimized. Design-related failures are failures due to:

- Incorrect or ambiguous requirements
- Deficiency in accounting for human use; lack of error-proofing
- Incorrect implementation of the design in meeting requirements
- Unspecified parameters in the design that should have been specified to ensure that the design works correctly
- Inherent design flaws that should have been found during design analyses, such as FMECA or verification testing
- High electrical or mechanical stress conditions which are beyond the strength of the design (which exceed design derating guidelines and manufacturer ratings)
- Design process weaknesses: lack of a robust review subprocess
- Probabilistic pattern failures and systematic failures that are random isolated incidences related to design weaknesses

How to Eliminate or Avoid Single-Point Failures

Single-point failures are addressed by assessing the corresponding criticality or RPN parameters. Once the RPN or criticality is calculated for a particular single-point failure, a design change is considered based on the relative ranking of the failure. If the failure is ranked high, a design change may be warranted. The design change may take several forms. There are instances when the results of FMECA conclude the need for design changes to improve reliability, in which the improvement in design involves changes in the system architecture. An example of a design improvement related to the system architecture is the enhanced capability to allow for fault-tolerant operation. Fault tolerance is the ability of a product or system to subdue the effects of a failure without causing a mission-critical effect. Fault tolerance is usually achieved by assuring that the system architecture includes redundant elements with active or passive data replication. This may include parallel redundant channels or signal paths. For redundant hardware elements in a subsystem, part of a fault-tolerant system architecture, a number of elements may fail while a minimum number of elements are required to operate to meet system performance requirements. This minimum number of elements compared to the total number of elements installed and configured in the system equates to a redundancy ratio used in system modeling. This type of system model may be referred to as an "*m out of n*" *system*.

When redundant elements involve multiple microprocessors, the architecture will include master–master or master–slave configurations for certain critical functions. The difference between master–master and master–slave configurations is the preset functional assignments of the processors. In a master–master configuration, both master processors are performing the functions together on the critical path. In a master–slave configuration, the master performs the preset assignment while the slave is held in reserve or backup, and usually does not perform a critical path function until promoted from slave to master when the master fails. Based on the system assets and the functional priorities, as a processor fails in a critical mission path involving a master–master or master–slave configuration, the processor assignments outside the critical path may be reassigned and a processor with a low-priority function may be promoted to serve in a master–master or master–slave configuration. Processor promotions and demotions occur continuously depending on critical-function failures, with noncritical functions taken off-line, critical function recovery, and failed processors brought back online to perform the noncritical functions. The fault-tolerant architecture offers flexibility to system designers who depend on certain critical functions to be available 100% of the time. Examples of digital and analog circuit design changes follow:

Digital Circuit Redesigned

- Built-in-test capability added to detect and isolate a failure
- Interruption of service (IOS) measured and high IOS alleviated by reducing the processing response times and decreasing service downtimes
- Test or fault coverage measured with causes of low probability of fault detection fixed
- Health status monitoring and prognostics capability added
- Checkpoints added
- Spares allocation schemes added (e.g., to avoid stack overflows or memory leakage failures)
- Fail-over processing added (e.g., master–slave processing)
- Asset synchronizing added
- Failure recovery mechanisms and redundancy schemes added

Analog Circuit Redesigned

- Design allocation for lower power use
- Use of components with lower power dissipation ratings
- Use of redundant circuits or channels
- Part ratings increase to design for more margin through lower stress percentages
- Use of capacitors that are open-mode style so that an open circuit will develop immediately when a capacitor shorts and current is drawn through the part

Probabilistic pattern failures and systematic failures that are random isolated incidences related to design weaknesses may need further discussion. Examples of these types of design failures are intermittent random failure events such as race conditions related to static or dynamic hardware or software timing, and incorrect use of shared data or global variables. These design-related failures include defects in specifications and requirements. One type of timing failure mode, the race condition, may be prevented with properly worded requirements, such as synchronizing timed events and controlling the use of processor interrupts that occur asynchronously. If functional timing requirements and interface requirements are specified properly, timing and race conditions can be eliminated.

PRINCIPLES OF FMECA-MA

Failure modes and effects analysis and maintainability analysis (FMECA-MA) add value to the design process by considering maintenance actions (preventive and corrective maintenance), fault diagnostics capability, and prognostics and health management (PHM) capability. Maintainability analysis is carried out during FMECA to analyze the design beyond the failure detection capabilities, to determine maintenance design features or repair actions that may contribute to the high or poor reliability of a system or product design. Maintainability analysis considers the following in conducting FMECA:

- Fault tolerance capability (redundancy, hot/warm/cold spares, active and passive replication, fail-over and fault recovery)
- Error detection and correction coding and automated schemes
- BIT effectiveness and ambiguity groups
- Fault detection probability and fault isolation probability: test coverage analysis and testability analysis
- Condition-based maintenance capability, including PHM and reliability-centered maintenance

During the execution of D-FMECA, the analyst must identify all the causes of failure for a particular failure mode. Failure modes include one or many failure symptoms. Failure symptoms are the characteristics of the failure that can define the failure on various levels, such as physical, electrical, mechanical, molecular, or atomic. Failure symptoms are failure effects at higher levels in the configuration of the system or product. There is a multiple one-to-many relationship in this analysis. For each failure identified (failure to meet specification or process), there are one or more failure modes. For each failure mode, there are one or several failure causes.

The Difference Between Process FMECA and Design FMECA

Process FMECA (P-FMECA), discussed in Chapter 6, is FMECA of the failure modes and failure causes related to a manufacturing or design process. D-FMECA is FMECA of the failure modes and failure causes related to a system or product design.

DESIGN FMECA APPROACHES

There are three approaches to D-FMECA:

1. In the *functional approach*, each item is analyzed by the required functions, operating modes, or outputs that it generates. The failure modes of a system are analyzed for specification and requirement ambiguities and hardware or software defects that allow a high potential for system faults-due to a lack of fault-tolerant design architecture. A functional block diagram is created to illustrate the operation and interrelationships among functional entities of the system and lower-level blocks as defined in engineering data, specifications, or schematics. Since this FMECA approach is highly dependent on complete and accurate product- or system-level requirements, functional FMECA may also be called requirements FMECA system design FMECA, architecture FMECA, or top-level FMECA.

2. In the *hardware approach*, all predictable potential failure modes of assemblies are identified and described. Each of the part- or component-level failure modes and failure mechanisms is analyzed to determine the effects at the next-higher indenture levels and product or system levels. Actual failure analysis data that identify the physics-of-failure mechanisms are useful in providing empirical data to replace the analytical data in the FMECA in terms of failure causes, failure modes, and failure effects. Hardware FMECA may also be called electrical design FMECA or mechanical design FMECA; these types of FMECAs are piece part or component bottom-up FMECAs. Hardware FMECA for assemblies may be carried out at the board or box level.

3. In the *software approach*, software design functions are analyzed. Software design FMECA includes analysis of software components, configuration items, modules and blocks analyzed during code walk-throughs, and code reviews to determine the potential of failure modes such as static and dynamic timing issues and race conditions caused by probabilistic failure mechanisms that could lead to system or product effects. Software FMECA is very similar to functional FMECA and is performed using a top-down approach. Further details of the software D-FMECA process are described in Chapter 7.

D-FMECA is begin early in the design process when the design specifications are written, before drawings, schematics, and parts lists are created. This is the type of D-FMECA known as functional D-FMECA. This D-FMECA is done from the top-level requirements down to ensure that the design requirements will incorporate features to handle mission-critical failure modes and to mitigate their effects. Fault-tolerant design capabilities and system sparing are the most common architecture approaches to handling mission-critical failure modes and to mitigate their effects. The concept of redundancy is the easiest fault tolerance implementation, but may be the most costly. The cost of redundancy depends on how much of the redundant capability is applied in spare mode. In spare mode, the redundant capability may only be exercised when a failure occurs or during peak operation, not during normal operation. This type of spare concept is called *cold spare*. By contrast, the concept of *warm* or *hot spare* may be applied for normal operating conditions, where the spare capability shares load and balances the stresses of the system and is not part of the redundant capability that is initiated upon occurrence of a failure during normal operation. Also, the cost depends on how much parallel capability is needed at the various hardware and software configuration levels. When employing redundancy, consider whether the redundancy should be implemented using hot spares, warm spares, or cold spares. In this case, the term *spares* refers to redundant elements of the design that are configured as powered-on elements operating in backup mode (hot spares), powered-on in standby mode (warm spares), or powered-off until needed or not connected to power (cold spares). Also, consider the redundancy in a master–master or master–slave configuration, and if active replication or passive replication is needed. Other costs associated with redundancy are added weight, space, and power usage. Besides redundancy, other considerations for design improvements during FMECA are part quality, new design/application risks, and design margin.

After an initial top-down functional D-FMECA, the next D-FMECA that may be performed is a hardware D-FMECA, which is performed when the drawings, schematics, and parts lists are created, but prior to building production hardware. This phase in design is the prototype phase, when hardware D-FMECA is carried out. In this type of D-FMECA, all parts are analyzed, looking at part failure modes involving functions and bus interfaces, and failure modes on pins, such as opens, shorts, and low impedances for analog devices, and stuck-at-one or stuck-at-zero states for digital devices.

This type of D-FMECA uses bottom-up system analysis. For the hardware D-FMECA approach, it begins at the lowest level of the system hierarchy. Only system and hardware D-FMECAs are described in this chapter; software design FMECAs are described in detail in Chapter 7. The failure modes of hardware D-FMECA include component or piece part–level failure modes and their failure causes. The system D-FMECA traces the effects of failure modes up the system hierarchy to determine the next-higher effects and the end effect on system performance. Hardware D-FMECA is typically implemented at the assembly or circuit card level, but may also be implemented at lower levels of hardware, such as modules and complex electrical or mechanical components.

EXAMPLE OF A DESIGN FMECA PROCESS

1. Start at the circuit card or assembly level to document all failure modes related to the component or piece part level, such as a digital integrated circuit (IC) on a circuit card assembly (CCA).
2. List and number all failure modes.
3. Consider the causes of part type failures, focusing on the pins of the IC and the functions of the device. Each pin could have an open or a short, high or low impedance, or low-voltage or high-current leakage.

4. List the causes of failure for each failure mode. Each failure mode could have several causes, such as a high-temperature failure mode caused by power dissipation, an open bond wire (in a dual-wire bond application for current handling), or delamination of the die substrate from the die paddle.
5. List the failure effects for each failure cause. Separate failure effects by the hardware indenture levels, starting with the CCA and working up the configuration levels to the system level for the end-item or end-user effects.
6. Assess the probability of occurrence, detection method, and severity for each failure effect.
7. Calculate the criticality for each failure effect in terms of the RPN.
8. Determine compensating provisions, safety issues, or maintenance actions for each that could change the criticality, increasing or decreasing the criticality calculation results or the RPN.
9. Enter all FMECA data into a database for easy data retrieval and reporting.
10. Sort by RPN or criticality.

After the failure modes for a particular design are found, steps should be taken to mitigate the failure effects and manage the failure correction process. The first is to verify failure modes and failure causes in the FMECA using failure reporting, analysis, and corrective action system (FRACAS) data. When test or field data are not available, the second choice is engineering analyses, failure mechanism modeling, durability analysis and models, or physics-of-failure models. Failure mechanism models and the like exist for many failure modes and failure causes. The failure mechanisms that are possible for electronic assemblies over their lifetime are associated with fatigue, corrosion, diffusion, wearout, fracture, and many other types of failure mechanisms. These can be estimated through engineering physics modeling and analysis.

Four of the most important reasons for performing an FMECA are to improve the design reliability through design changes, to learn about the design for documenting why a failure might occur and how the design detects and reacts to a failure, to conduct a system safety analysis, and for risk assessment. Performing FMECA alone does not improve the reliability of the design. The reliability is improved only when design changes are implemented that avoid failure modes, minimize the probability of failure mode occurrences, lessen the severity of failure effects, and/or alter the architecture to incorporate fault tolerance features, which may include functionality and circuit redundancy and/or increase the capability and efficiency of the design to detect, isolate,

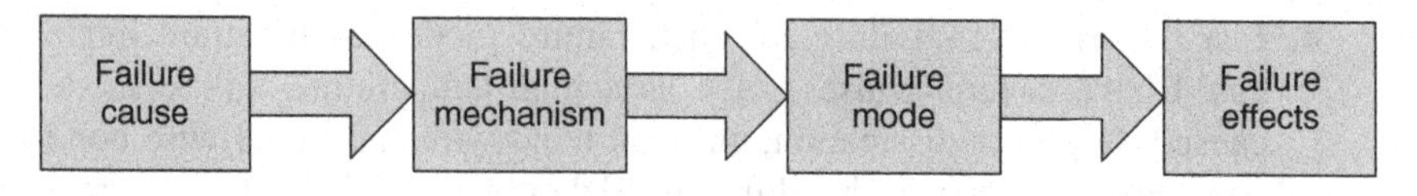

Figure 1 Failure modes and effects relationship diagram.

and recover from failure modes. Figure 1 illustrates the relationships among failure cause, failure mechanism, failure mode, and failure effect:

Failure cause: the situation or conditions during the design, production, or application of a device, product, or system which explain the underlying cause of a problem that led to a failure mechanism.

Failure mechanism: the physical results of a root cause failure of a device, product, or system that led to a failure mode.

Failure mode: an unwanted state or function of a device, product, or system that is either inherent or induced, and explains its inability to function as desired or demonstrates a gradual degradation in performance that could lead to a failure effect.

Failure effects: the results of a failure mode on the higher levels of assembly up to the end-item level or integrated functions adjacent to the device, product, or system, which has failed to perform its function, has reached an unacceptably high probability of failure, or has an increasing probability of failure.

As defined in IEEE 1413.1 [6], the time-related characteristics of a failure mode falls into one of the four following categories:

1. Items that exhibit constant rates of failure. Such products include electronic and electrical components.
2. Items that exhibit degradation over time, "wearout" failures. Such products include electromechanical and mechanical components and pyrotechnic devices.
3. Items that exhibit decreasing rates of failure over time, "early life" failures. Such products include low-quality parts and systems that are subject to a reliability growth program.
4. Items that exhibit combinations of failure types 1 to 3.

An inherent failure mode is due to characteristics of the product that cannot be eliminated or controlled to avoid failure. An example of an inherent failure mode is a dead flashlight battery, because a battery has a finite life. On the other hand, an induced failure mode is due to characteristics of the product or external factors that can be either eliminated or controlled at some point in the life cycle of the product to avoid failure during operation. If the opportunity to eliminate or control a failure mode is not exercised, whether

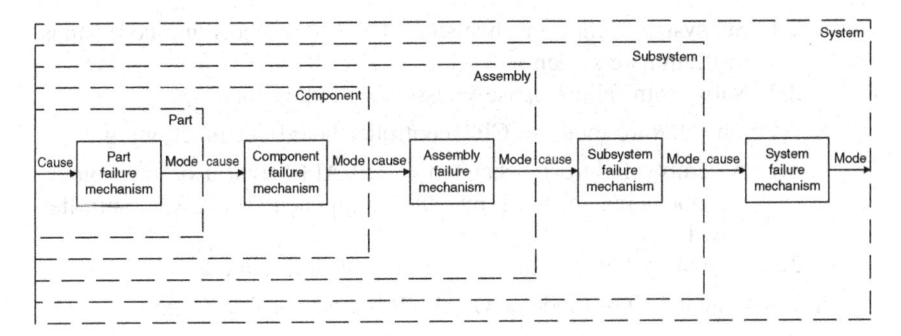

Figure 2 Failure flow diagram. (From [6].)

voluntarily or unknowingly, the failure mode is classified as process-induced. Process-induced failure modes can be identified and mitigated by means of verifiable compensating provisions, such as design changes, special inspections, tests, or controls, or by operational corrective actions such as replacement, reconfiguration, repair, or preventive maintenance.

Failure mode data are often represented in the form of a Pareto chart or histogram, which plots the failure modes by decreasing frequency of the occurrence of failures over a certain time period, or decreasing RPNs. In addition to a failure mode being either inherent or induced, it is also either potential or actual. A potential failure mode is the capability of a product to exhibit a particular mode of failure. An actual failure mode is the affectivity of a product to exhibit a particular mode of failure. For a potential failure mode to become an actual failure mode, a failure cause must act as the catalyst through evidence of failure mechanisms. In general, the exhibited failure mode at one level of the system becomes the failure cause for the next-higher level. This bottom-up failure flow process applies all the way to the system level, as illustrated in Figure 2. The same process flow can be used in the opposite direction, with the top-down approach. For every observable failure mode there is an initiating failure mechanism or process that is itself initiated by a root cause. The figure shows the steps that could be followed in D-FMECA using available diagnostics to analyze a computer monitor failure using the top-down approach to illustrate the failure flow process.

The top-down system failure process flow example for a computer monitor failure is described below.

1. System failure mode = computer not functioning
   - **1.1.** System failure mechanism = operator cannot see images on computer monitor
   - **1.2.** System failure cause = subsystem failure mode
2. Subsystem failure mode = computer monitor not functioning

   - **2.1.** Subsystem failure mechanism = CRT is not receiving commands to illuminate screen
   - **2.2.** Subsystem failure cause = assembly failure mode
3. Assembly failure mode = CRT controller board not functioning
   - **3.1.** Assembly failure mechanism = OP AMP hybrid output signals are not being received by other components on CRT controller board
   - **3.2.** Assembly failure cause = component failure mode
4. Component failure mode = OP AMP hybrid not functioning
   - **4.1.** Component mechanism = PNP transistor output signals are not being received by other parts in OP AMP hybrid
   - **4.2.** Component failure cause = part failure mode
5. Part failure mode = PNP transistor not functioning
   - **5.1.** Part failure mechanism = PNP transistor base pin is not receiving input signals due to an open solder connection
   - **5.2.** Part failure cause = manufacturing process failure mode
6. Manufacturing process failure mode = solder connection to PNP transistor base pin is open
   - **6.1.** Manufacturing process failure mechanism = insufficient solder flow applied to PNP transistor base pin
   - **6.2.** Manufacturing process failure cause = inadequate testing, defective soldering equipment, faulty soldering procedures, or human error

The failure mode data are used to determine the adjustments needed for estimating the reliability of a product of interest that is being compared to an in-service product for which the failure modes and failure rates are known. For example, if there were a new design that eliminates most of the design-process-induced failure modes present in an in-service product, the failure rate predicted for the new design would be less than that of the in-service product.

Examples of failure causes for inherent and process-induced failure modes are shown below. Each of these failure causes may act as the catalyst for turning a potential failure mode into an actual failure mode that leads to the functional failure of a product. For example, low component quality or chemical reactions may lead to early life failure. Design rule violations or overstressed parts may lead to intermittent operation. Uninspected workmanship or test equipment faults may lead to inhibiting desired functions. Excessive operating temperature or humidity may lead to electrical open or short conditions. Finally, incorrectly calibrated instruments or operator errors may lead to initiating unwanted functions. Therefore, it is safe to say that a system-level failure mode may result from a hardware fault, a software fault, customer abuse, a design error, a manufacturing error, or many other possible causes.

Inherent Failure Causes

- Chemical reactions
- Specified design life exceeded
- Wearout
- Expendables exhausted (e.g., spent fuel, oil quality/cleanliness, filter clog)
- Fatigue threshold exceeded

Design Process–Induced Failure Causes

- Design rule violations
- Design errors
  - Overstressed parts
  - Timing faults
  - Reverse current paths
  - Mechanical interference
  - Software coding errors
- Documentation errors
- Nontested failures
- Internal electromagnetic threshold exceeded

Manufacturing Process–Induced Failure Causes

- Uninspected workmanship
- Test equipment faults
- Handling errors
- Test procedure errors
- Assembly faults
- Rework faults

Environmentally Induced Failure Causes

- Excessive operating temperature
- Humidity
- External electromagnetic threshold exceeded
- Foreign object damage
- Vibration

Operator- and Maintainer-Induced Failure Causes

- Incorrectly calibrated instruments
- Operator errors

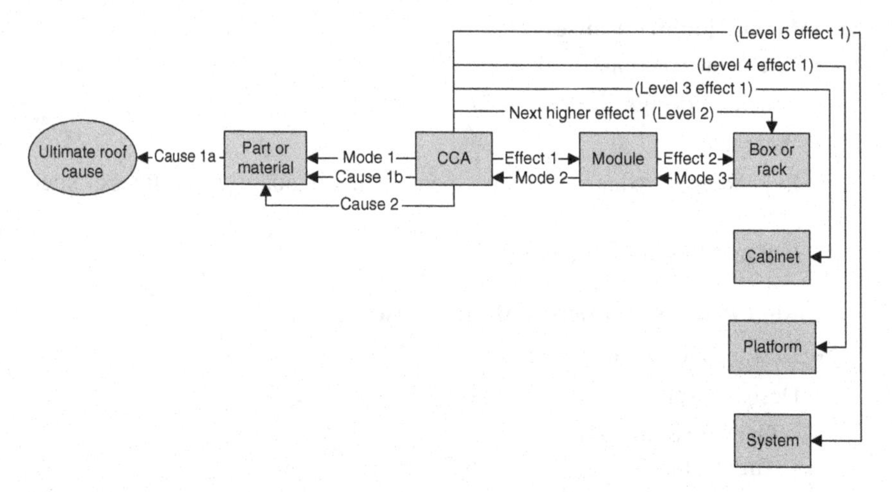

Figure 3 CCA failure flow diagram. (From [6].)

- Maintenance errors
- False system operating status
- Unnecessary maintenance actions

The failure mode at one functional level is a failure effect at another functional level. Let's look at Figure 3 and the FMECA failure modes, failure causes, and failure effects from the perspective of the CCA. The local effect (level 1 effect) of a CCA failure mode is located at the module level. The next-higher failure effects for this same failure mode occur from level 2 at the box or rack level and flow up to level 5 (system effect). FMECA will prove to be value-added during the design development process when its outputs are useful to design engineers and other engineering disciplines that support the development process.

An example of a failure cause distribution compiled from the results of actual test data collected from line replaceable unit (LRU) failures is provided in Figure 4. The numbers shown in the figure are percentages associated with each failure cause that sums up to 100%. Figure 5 illustrates a failure cause Pareto diagram that should be prepared from the actual failure data for preparing priorities for a corrective action plan to resolve failure cause issues. The numbers represent actual failure counts for each failure cause type. To keep the math simple, the failure count was normalized to 100. The failure counts in Figure 5 correlate with the failure percentages in Figure 4.

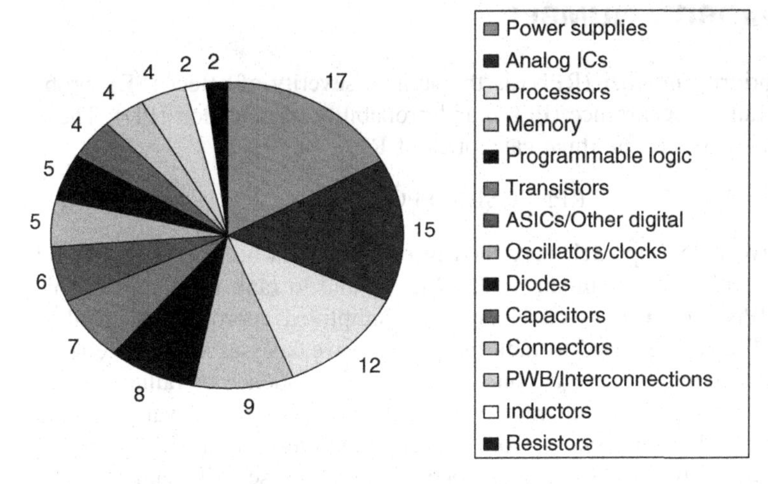

Figure 4 LRU failure cause distribution.

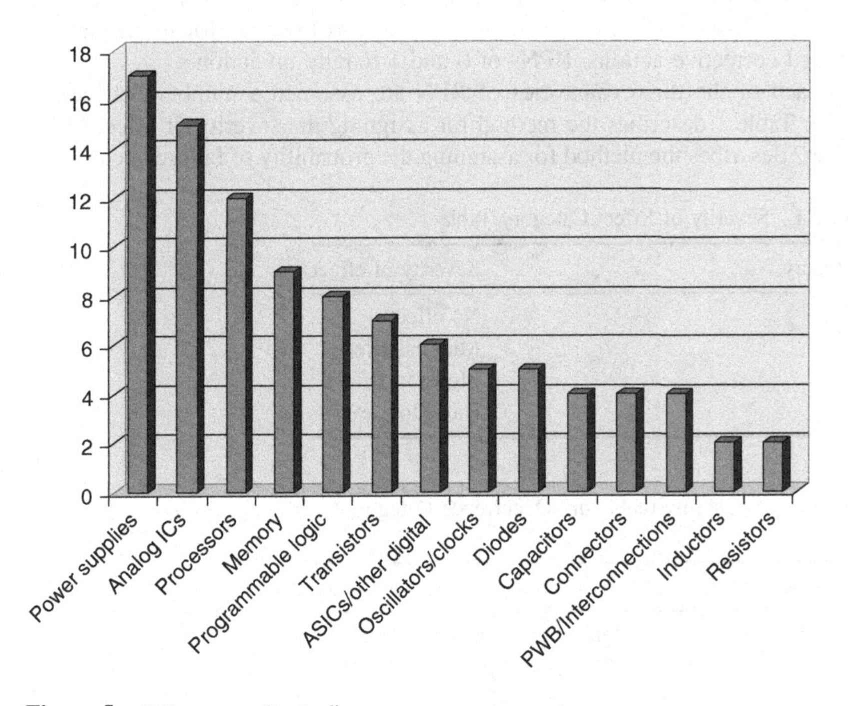

Figure 5 Failure cause Pareto diagram.

RISK PRIORITY NUMBER

The risk priority number (RPN) is the sum of severity of effect (SE), probability of failure occurrence (PFO), and probability of detection (PD). These three terms represent the three categories of RPN:

$$\text{RPN} = \text{SE} + \text{PFO} + \text{PD} \tag{1}$$

The additive RPN approach described in equation 1 is not the only method used to prioritize failure modes in D-FMECA and to rank the failures. This additive RPN method was developed as a simplified approach to calculate the RPN. The RPN is used to rank the potential weaknesses so that the team can consider more effective actions to reduce the incidence of failure modes for critical and significant characteristics and to reduce process variation and accordingly make the process more robust. Regardless of the RPN, special attention should be paid to failure modes with a high severity number. The threshold for the RPN could be assigned a score of 6 out of a possible 9, so that any failure mode scored between 7 and 9 will be corrected immediately with a design change. Moderate RPN failure modes, such as 5 to 6, are planned corrections later in the program. RPNs between 2 and 4 are included on a watch list for collection of further data to support the need for immediate or planned corrective actions. RPNs of 0 and 1 require no action.

Each of the three categories of RPN are assigned a number between 0 and 3. Table 1 describes the method for assigning the severity of effect (SE), Table 2 describes the method for assigning the probability of failure occurrence

Table 1 Severity of Effect Category Table

Category	Severity of effect
0	No effect
1	Minimal effect
2	Moderate effect
3	Hazardous effect

Table 2 Probability of Failure Occurrence Category Table

Category	Probability of failure occurrence
0	Remote probability
1	Slight probability
2	Moderate probability
3	High probability

Table 3 Probability of Failure Detection Category Table

Category	Probability of detection
0	High probability to detect
1	Moderate probability to detect
2	Sllight probability to detect
3	Low probability to detect

Table 4 RPN Table

Categories of RPN					
SE	PFO	PD	RPN	Type of action recommended	Action taken
0	0	0	0	N/A	None
0	0	1	1	No improvements necessary for RPN 1	None
0	1	1	2	All RPN 2; add to the watch list; monitor for trends	Monitor
1	1	1	3	Same as RPN 2, and conduct analysis and collect data; determine when to plan a fix	Analysis
1	2	1	4	Same as RPN 3; frequent failures, collect more data for planning a fix	Analysis
1	1	2	4	Same as RPN 3, except poor detectability; wait for more data prior to planning a fix	Analysis
1	2	2	5	Frequent failures but not severe effects; add to plan for change in the next one or two design cycles	Plan
2	2	2	6	RPN 6 with severe effects; highest priority for design change under the next revision cycle	Plan
3	2	2	7	RPN 7, 8, and 9; design change immediately to avoid severe failure effects	Yes
2	3	2	7	RPN 7, 8, and 9; design change immediately to avoid excessive customer issues	Yes

(PFO), Table 3 describes the method for assigning the probability of detection (PD), and Table 4 illustrates how to determine recommended actions based on the RPN score and the categories of RPN.

RPN Ranking

A Pareto chart is a preferred method to illustrate results for stack-ranking the design change priorities and reveal the most critical entries, which are

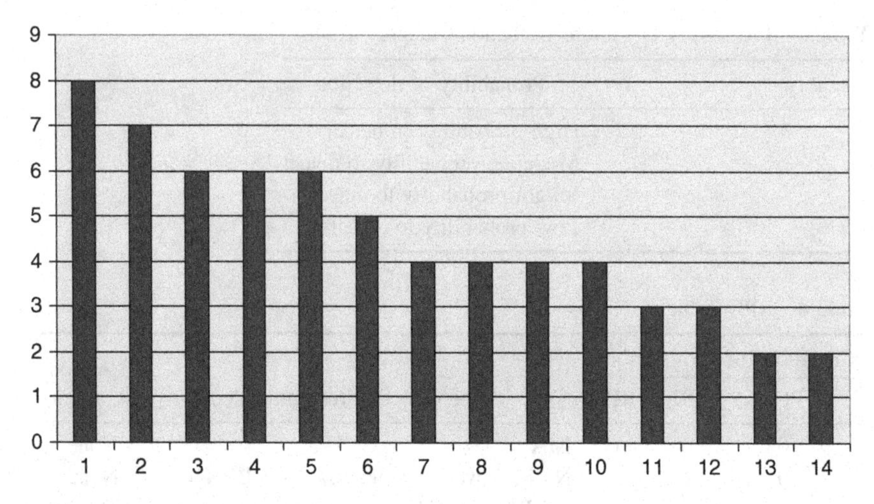

Figure 6 RPN Pareto chart.

those failure modes to focus on that warrant design investment to prevent further life-cycle cost exposures. This ranking of RPNs with associated failure modes and failure causes provides a method to organize and plan improvements to the design, process, or test and ultimately, the system or product reliability. Figure 6 ranks a few sample RPN values from the failure cause analysis Pareto shown in Figure 5. Figure 6 shows a method to calculate the RPNs for 10 example failure causes using RPNs ranging from zero to 7.

Actions Recommended

Emphasis should be placed on developing actions aimed at preventing or eliminating root causes. Detection controls may be valuable in preventing failure modes, but they can be costly and inefficient. Reliance on random sampling and 100% inspection should be avoided since they are inefficient and costly. Identify actions that stabilize and improve the process. FMECA provides a method to build a failure management capability during the early development phase of a program that offers value through detection of failure modes and improvements in the design prior to hardware prototype, production build, or software coding. FMECA data for fault management represent a continuous and evolving process that should be used and maintained throughout the system or product life cycle. FMECA leverages failure data from FRACAS data from the same product, a similar product or program, or reliability prediction data combined with FRACAS data. Failure modes reported in testing and in customer applications validate the FMECA conclusions, or result in changes to the FMECA to improve accuracy. RPN scores that are slightly below the threshold for immediate design improvements will be deferred to later in a program when a large change order is planned to upgrade and enhance the functionality of the design. At this point, design changes to correct bugs, errors, and defects deemed moderate priority earlier in the program can also be implemented. This may be possible through the execution of spiral system development on a program. During periodic intervals of time (e.g., semiannually), the design is refreshed with the latest technology, component obsolescence issues are resolved, new customer functionality is added, and increasingly higher-priority failure modes are corrected.

Action Results and Revised RPN

One or more follow-up meetings must be scheduled to ensure that all actions recommended have been implemented and have achieved the desired results. Enter a brief description of the actions taken, the effectiveness of the actions, and the actual completion date. After the actions are completed, the FMECA team must revise the ratings for severity, occurrence, and detection and recalculate the RPNs. Review the new RPNs to decide if further process actions are necessary.

Most system problems today arise from system management problems. These problems may be in the form of requirements deficiencies, communication problems, and interface management problems. Appropriate design features for situational awareness decreases the design uncertainty and improves operational reliability. FMECA aids designers in creating these situational awareness design features by enhancing the reliability through the use of collaborative tools such as FMECA with decision-maker and stakeholder involvement. In many instances, I have seen FMECA engineers working apart from the design engineers in doing FMECA. Engineers performing FMECA usually perform their work in a vacuum and do not communicate with their designer counterparts.

They usually do this for good reasons (e.g., did not slow down the design process while performing their analysis in parallel). The FMECA intent was to produce an artifact. However, the FMECA process has a positive impact on the design if designers are involved. When performing FMECA, designers will change their design based on the realization that FMECA brought out ideas that they had not considered earlier. FMECA has the most value to the design when this occurs, such as the timely interaction of stakeholders that occurs at a phase in the development cycle when changes can be made to improve the reliability of the design.

FINAL THOUGHTS

To truly affect design for reliability, influencing designers to improve their design using the FMECA methods must occur. A FMECA analyst should conduct open dialogue sessions with the system or product designers and stakeholders. The analyst should act as a facilitator to brainstorm ideas and capture all thoughts without passing judgment. The analyst should be able to gain valuable design information and data from the sessions to properly complete the FMECA; maintain points of contact for peer reviews, visibility, and follow-up actions; engage stakeholders, and promote teamwork through use of collaborative tools.

REFERENCES

[1] *Potential Failure Modes and Effects Analysis in Design (Design FMEA) and Potential Failure Modes and Effects Analysis in Manufacturing and Assembly Processes (Process FMEA) Reference Manual*, SAE J1739, Society of Automotive Engineers International, Warrendale, PA, July 1994.

[2] *Potential Failure Mode and Effects Analysis*, 3rd ed., FMEA-3, Automotive Industry Action Group, Southfield, MI, July 2001.

[3] *Recommended Failure Modes and Effects Analysis Practices for Non-Automobile Applications*, SAE ARP5580, Society of Automation Engineers International, Warrendale, PA, July 2001.

[4] *Analysis Techniques for System Reliability: Procedure for Failure Mode and Effects Analysis (FMEA)*, IEC 60812, International Electrotechnical Commission, Geneva, 2006.

[5] *The Reliability Analysis Center Failure Mode Distribution Book*, RAC, Utica, NY, 1997.

[6] *IEEE Guideline for Reliability Prediction*, IEEE-1413.1-2003. IEEE, Piscataway, NJ, 2003.

[7] *DoD Guide for Achieving Reliability, Availability, and Maintainability*, U.S. Department of Defense, Washington, DC, 2005.

[8] Stamatis, D. H., *Failure Mode and Effect Analysis: FMEA from Theory to Execution*, American Society for Quality, Milwaukee, WI, 2003.

[9] McDermott, R. E., Mikulak, R. J., and Beauregard, M. R., *The Basics of FMEA*, 2nd ed., CRC Press, Boca Raton, FL, 2008.

Chapter 6

Process Failure Modes, Effects, and Criticality Analysis

Joseph A. Childs

INTRODUCTION

As discussed in Chapter 5, failure modes, effects, and criticality analysis (FMECA) is a tool used for addressing risk of failure. In the case of design FMECA (D-FMECA), it is extremely valuable in reviewing the design aspects of a product—from the standpoint of the ability to investigate possible issues for a product meeting its design requirements. Just as D-FMECA focuses on possible design failure modes at multiple levels of hardware or software structure, P-FMECA is a detailed study, focused on manufacturing and test processes and steps, required to build reliable products. The format used in this chapter is different from that discussed previously for D-FMECA, but the considerations are the same: Risks: how serious the potential event is, the chances that it will happen, and how users know about it.

PRINCIPLES OF P-FMECA

P-FMECA includes a review of materials, parts, manufacturing processes, tools and equipment, inspection methods, human errors, and documentation. P-FMECA reviews the build, inspection effectiveness, and test aspects of the

Design for Reliability, First Edition. Edited by Dev Raheja, Louis J. Gullo.

Product or Product Line __________ __________ **P-FMECA** Team Members __________

Process or Sub-Process __________ Revision Date __________

Item #	Process name and description	Process step	Failure mode	Failure effect	SEV	Potential causes	OCC	Verification method	DET	RPN	Recommended Actions

Figure 1 P-FMECA form.

product for possible risks of process step failures at the multiple levels of processes, including the severity of problems after the product is in the customer's hands.

The P-FMECA process is broken down into subprocesses and lower-level steps. For each step, possible problems, their likelihood of occurring, and their ability to be detected before extensive consequences are explored for each step. The impacts to the customer, as well as to subsequent subprocesses, are considered. Throughout this process, a matrix is used to help define the scope of effort, to track the progress of the analysis, to document the rationale of each entry, and to facilitate quantification of the results. Figure 1 is an example of one type of matrix used for such an analysis. This quantification process, along with the rationale, is extremely useful, when working to prioritize possible improvement recommendations. The details of how to use this form as a tool to complete the analysis are discussed in subsequent paragraphs.

USE OF P-FMECA

Why P-FMECA Is Covered in a Design Reliability Book

Review of failure data in many programs indicates that many of the causes of product failures were related to processes, especially the manufacturing and test processes. These tie directly to the design function responsibilities, especially as they relate to reliability. The following considerations should be taken into account:

- The design engineer knows more than anyone how a system or piece of equipment works: what is important to its operation and what is not. Often, the design engineer plays a significant role in making decisions that drive the manufacturing process.
- The designer selects parts, defines layout and materials, and makes a myriad of decisions that define how a design is to be built and tested. For example, if a new material or component is required to meet requirements, this could affect the thermal profile used to attach

the part. Often, issues with manufacturing or testing are avoided or resolved by means of the design. Design decisions must be selected considering cost, schedule, and performance.

- P-FMECA is best performed with a team of process stakeholders, including staff from manufacturing, test engineering (or technicians), maintenance (when that is a key issue), reliability engineering, quality engineering, and design engineering. This diversity provides a variety of viewpoints to consider for various issues.

An interesting example used in P-FMECA training is to consider a fire in a building. One question is: How would this affect us? Many different and interesting answers might ensue. What would cause this to occur? Almost without exception, design engineers would provide issues with assembly, installation, maintenance, and inspection, while maintenance engineers and technicians would bring up possible design problems.

The point is that *the fidelity of FMECA is improved by having multiple viewpoints*. This is not without cost: The more personnel who are involved in the analysis, the more discussion takes place, and this takes time. However, if guided properly, such discussion is enlightening, and the effort yields more complete and useful results. For example, a design may include an elegant means of alignment of piece in an assembly. However, if the operator must perform the alignment by "touch" or by "eyeballing," the results almost certainly vary with operator experience, state of mind, and physical ability. Adding a fixture to assure less variability in results could be an invaluable improvement to the process at little cost. Upon completion of successful P-FMECA, some or all of the following benefits are realized:

- Improved reliability
- Improved quality and less variability (higher yields, diminished schedule, and cost risks)
- Improved safety in the manufacturing processes
- Enhanced communication between the process owners
- Improved understanding of processes and interfaces by participants

Overall Benefits of P-FMECA

The benefits of FMECA are the same as those for its design counterparts, but for processes. The total process can be considered analogous to a new design to undergo D-FMECA. Then the subprocesses and lower-level single steps are like the assemblies and the piece parts. The idea is the same: to break down the top entity to consider possible issues and their causes, judge their effects, and improve their design. The analysis results in a thorough review of the process with a prioritized list of possible problems and an improvement plan or recommendation.

Experience shows that P-FMECA is a tool not only to identify process concerns but also to *motivate those who do the follow-up*. After one particularly detailed D-FMECA process, a project design engineer said, "I think this is a wonderful tool. I never knew so much about this system . . . ." After a pause he went on to say, "But I *never* want to do one again . . . *Ever*!" When asked why, he responded, "Because, since I've done this, I haven't had a good night's *sleep*! I've never thought about so many things that could go wrong!" Needless to say, this designer was quite anxious to complete the improvement actions we developed over the course of that analysis. The same results have been observed from P-FMECA as well over the course of performing many such analyses.

One other important aspect of FMECA: Each result is prioritized with corresponding rationale. This allows the team to consult with management as to which items deserve highest priority. This, combined with cost estimates of recommendations, provides a logical decision-making tool.

WHAT IS REQUIRED BEFORE STARTING

Next we discuss key support considerations when performing P-FMECA. Accounting for these points is needed to assure a successful analysis—one that points out key risk, how such risks can be mitigated, and why they are recommended.

Management Support

Since P-FMECA is complex, it is best performed by a team. This provides an optimal environment for uncovering possible concerns and improvements. Management must support the time, money, and resources it takes to perform this effort. For a very complex, critical process, the process is longer, but the results are more valuable as well.

Cross-Functional Team

Earlier it was noted that using a team from multiple disciplines is quite useful for this type of analysis. This helps to assure multiple viewpoints and can sharpen the focus of the results to account for various possible sources and interactions of issues and their causes. Functions that typically use P-FMECA are:

- Reliability engineer

- Manufacturing engineer
- Design engineer
- Quality engineer
- Senior line operator or supervisor

Planning

With the team established and a management go-ahead, the analysis should be well planned regarding the schedule, budget for the analysis itself, ground rules for the participants, analysis location, and meeting frequency. One key element in planning is scope. It is critical that the ground rules include the processes and subprocess to be covered, and what areas will not be covered. This will help the participants stay on track and help avoid additional unplanned effort

PERFORMING P-FMECA STEP BY STEP

Team Preparation

In selecting specific team members, it is important to assure that the team participants can understand the specific processes and steps to be performed, the tooling and fixturing involved, the design of the equipment undergoing the processes, and how the unit or system is to be used by the customer. The first step in performing P-FMECA is to list the processes and steps to be considered. In effect, this task actually translates the defined scope into the specific areas that will be covered. A process flow diagram, as well as the specific procedures and assembly drawings to be analyzed, are needed to perform P-FMECA. They should be kept on hand as reference material during the analysis. In fact, in many cases (where possible), examples of the actual hardware to be built, assembled, or otherwise processed should be available as well. Also, trips to the process lines to view the actual activities are extremely useful to assure that participants are aware of the details involved.

The ground rules of the effort should be established up front. For example, since each team will have its own point of view, it is valuable to settle on the criteria for ranking of specific factors (discussed further later in the chapter). A general description of the ordinal values is provided here. However, specific details and tailoring to unique aspects of a design or process will save much time later in the analysis. Specific meeting times, locations, and tabling of subjects not to be covered in P-FMECA can be covered at this time as well. A practice that has been found useful is to create a "parking lot" to collect

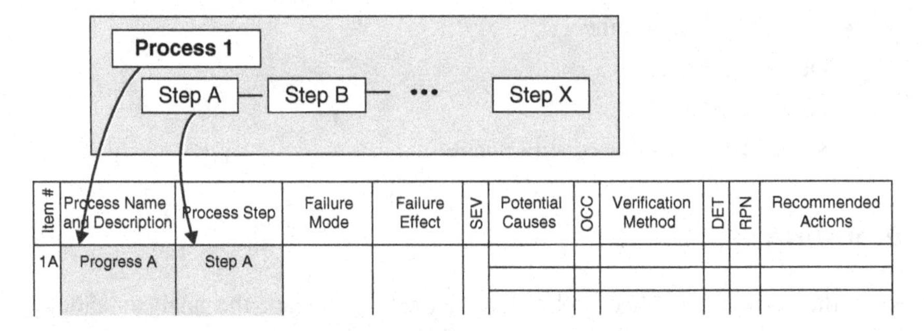

Figure 2 Defining the process and subprocesses to be analyzed.

subjects that although important or useful to discuss are outside the scope of the analysis at hand. These points can be covered at another time.

Defining the Processes and Subprocesses

Figure 2 shows the two columns "Process Name" and "Process Step" being completed. This is the first part of a P-FMECA effort. By completing the first two columns the team actually defines the scope of the analysis. That is, if the process steps are analyzed and there are no other additions or changes, this step actually defines the level of detail and which steps are covered in the analysis. It is sometimes necessary to include additional columns. This may be necessary, based on the way the process owner has defined the process to be studied. Also, in the course of P-FMECA, other processes may be identified as worthwhile to analyze, or another change may be considered necessary. These changes are expected and can be implemented as long as the team recognizes that this is a change in the scope of the analysis.

Failure Modes and Effects: The SEV Factor

When the analysis scope is defined, the steps are reviewed for how they might fail. In this context, a failure is when a step is missed, done incorrectly, or a defect can result. The idea is to delineate the list of possible ways in which mishaps can occur (failure modes) and look at the rest of the process to see how the entire process could be affected (failure effects). Figure 3 shows how these considerations fit in the P-FMECA form. Any failure could result in increased cost, loss of quality, schedule impact, and even harm to equipment or personnel. More specific ways in which a failure could affect the process are:

- Interruption of subsequent process steps or processes

FAILURE MODES AND EFFECTS – What could occur in this process step that could result in production or product issues? In what way?

Item #	Process Name and Description	Process Step	Failure Mode	Failure Effect	SEV	Potential Causes	OCC	Verification Method	DET	RPN	Recommended Actions
1A		Step A	Wrong Part	Wasted product, poor or no throughput; low yield							

Figure 3 Failure modes and their effects.

Table 1 Severity Factor Guideline Example

Severity of effect	SEV
Product or plant safety at risk; risk of noncompliance to government regulations	**10**
Major impact on ability to produce quality product on time; includes significant interference with subsequent steps or damage to equipment; could result in mission failure	**9**
Product defect, rejection, failure in in-spec storage/operational environments; disruption to subsequent process steps	**7–8**
Customer dissatisfaction, some degradation in performance, loss of margin, or delays in the process	**4–6**
Slight customer annoyance, slight deterioration in performance or margin, minor rework action or in-line delays	**2–3**
Little or no effect on product or subsequent steps	**1**

- Defects that could result in wasted material, rework labor, or schedule time to repair
- Unreliability or poor quality in the product shipped
- Unsafe situation on the line or in the field

The importance of the impact of these issues is the basis for each of the severity (SEV) factors defined. This is why each failure mode has a separate line and all the major possible effects are listed on that line. The SEV factor (value 1 to 10) is selected based on the most severe possible effect. Table 1 illustrates how such values can be guided by an objective listing of values. The more specific the rationale, the more consistent the analysis will be. Tables are provided as well for the likelihood of occurrence (Table 2) and the detectability of the occurrence (Table 3).

Table 2 Detection Factor Guideline Example

Likelihood of failure	Failure rate: Defect rate	Failure rate: Sigma	OCC
Failure is certain or almost inevitable	1 in 2		**10**
	1 in 8		**9**
Failure trend likely; process step not in statistical control; or SPC not used; similar steps with known problems; little/no experience with new tool or step	1 in 20		**8**
	1 in 40		**7**
Possible failure trend; process is in statistical control (CPK < 1.00); similar steps with occasional problems	1 in 80		**6**
	1 in 400		**5**
	1 in 1,000	$\sim +3\sigma$	**4**
Low likelihood of failure; step in statistical control (CPK > 1.00); similar steps with isolated occurrences	1 in 4,000	$\sim +3.5\sigma$	**3**
Very low likelihood of failure; in staticstical control (CPK > 1.33); rare occurrences in similar steps	1 in 20,000	$\sim +4\sigma$	**2**
Remote; no failure in similar steps (CPK > 1.67);	1 in 1,000,000	$\sim +5\sigma$	**1**

Table 3 Detection Factor[a] Guideline Example

Likelihood of detection	OCC
No means of detection; no process or equipment to find problem in time to affect outcome	**10**
Controls would probably not detect defect or failure; operator to perform self-inspection	**9**
Controls have poor chance of detecting defect; inspection alone to detect problem	**7–8**
Controls might detect defect; double inspection or inspection with equipment aids	**5–6**
Controls have good chance of detecting defect; process equipment detects presence of problem under most circumstances	**3–4**
Controls will almost certainly detect defect; process detects defect automatically	**1–2**

[a]Likelihood of detecting defect before next process step, or before product leaves the manufacturing or assembly station (i.e., detection that would lower the impact or likelihood of the event).

POTENTIAL CAUSES – What are possible causes that could cause this failure mode?

Item #	Process Name and Description	Process Step	Failure Mode	Failure Effect	SEV	Potential Causes	OCC	Verification Method	DET	RPN	Recommended Actions
1A		Step A	Wrong Part	Wasted product, poor or no throughput; low yield	9	Wrong marking					
						Unclear drawing or WI					
						Operator error – training					
						Operator error - too busy					

Figure 4 Potential causes.

Possible Causes: The OCC Factor

The next factor is intended to consider the possible ways in which the failure mode could occur (Figure 4). It is not uncommon for there to be several causes. An *immediate cause* is often defined as the step done incorrectly that would result in the problem. The *root cause* is found by asking the question: What would have caused that? This is asked many times until an *actionable cause* is determined. The point is to get to a reason for a possible problem that can ultimately be addressed and corrected. For example, it is not enough to state that the operator did something incorrectly. It is more important to look at what kinds of pressures are on the operator, how well trained the operators are, whether there are adequate instructions and pictures for the intended result, and whether the designs are possibilities for the tooling/fixturing or the product itself that could lower the likelihood of failure. A simplified fault tree (Figure 5) is a useful tool for helping to define the root causes, since it allows a pictorial way to view these types of questions and their answers. In this example, the fault tree is shown with no logic symbols. In most cases, the actions are all series—one of the lower-level causes will result in the next-higher-level event. This is not always the case. In instances where more than one contributing cause must occur, these must be shown with AND gates, implying two or more causes. This can be important, because the likelihood of two or more occurrences is much lower than that of just one occurrence. In general, for processes such as building, testing, and shipping, the causal relationships are serial in nature, so logic symbols are not needed. More detail about fault trees and their use in reliability problem solving is treated further.

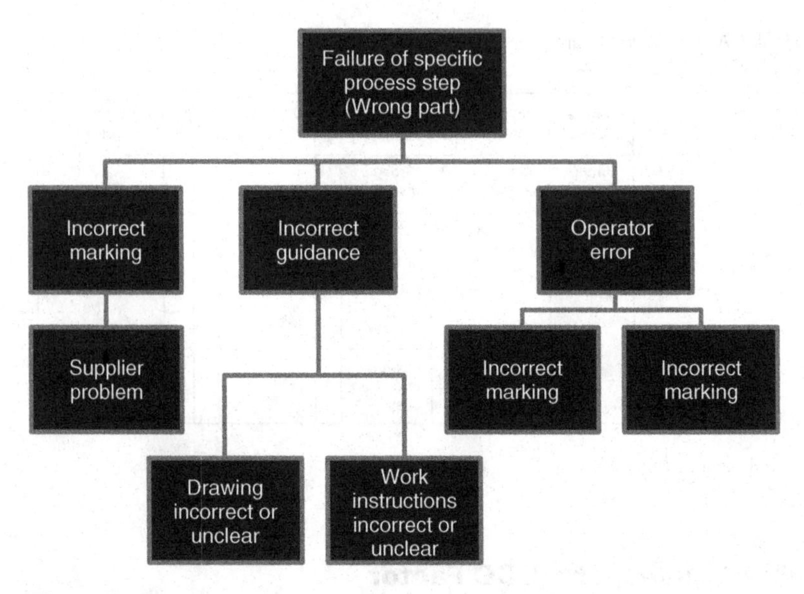

Figure 5 Simplified fault tree example: a tool for finding root causes.

Each cause listed is assigned an OCC factor. In all three factors, the values vary between 1 and 10. Table 2 is an example of how the values could be made objective and consistent. This particular example is especially useful if the process being analyzed is tracking defect data for statistical process control. In that case the sigma values represent a statistical method for estimating process issues in a quantifiable way. If these data are not available, other historical information or the judgment of the team must be used to arrive at a reasonable value for OCC.

Shown in Figure 6 are only individual causes, which are frequently found in process issues. However, if two or more causes were needed for the next-level cause or failure to occur, these would be considered in parallel. In practice, Boolean logic symbols would show when a single cause (OR) or multiple causes (AND) are needed. In Figure 6, no presence of logic symbols implies that only ORs are being used.

Verification Method: The DET Factor

The final factor to be determined is the detection (DET) factor. The main underlying idea is to gain insight into how well the process is set up to detect a failure, defect, or other flaw in a manner that would prevent a worsening result. Clearly, this and the SEV factors are interrelated, and the team should

VERIFICATION METHOD – How would the cause/mode be detected in time to avoid negative impact on the line or the product?

Item #	Process Name and Description	Process Step	Failure Mode	Failure Effect	SEV	Potential Causes	OCC	Verification Method	DET	RPN	Recommended Actions
1A	Procsss 1	Step A	Wrong Part		9	Wrong marking	2	Incoming & in-line QC inspection	6		
						Unclear drawing or WI	5	None	10		
						Operator error-training	2	Yearly training sessions, weekly group meetings and briefings	2		
						Operator error - too busy	7	None	10		

Figure 6 Verification method.

be flexible in its evaluation of them. Each cause has its own verification method(s) and DET values.

Questions to be addressed are:

1. How would this failure mode be detected based on the cause listed? Would the operator or others be able to understand what the problem is and how to address it in time to avoid worst-case results?
2. Are there multiple means for the failure mode to be detected? If so, all the methods of detection should be accounted for in determining the DET value.
3. Are the detection methods formalized as part of the process? Are they automatic or built in? Manual? Or occasional?

The DET factor, then, is answering the question: How likely is it that the cause could be detected in a time or place where the effects could be mitigated? If multiple means of detection are used, all those methods would be included in evaluating the DET rank value (Table 3).

Risk Priority Number

The risk priority factor (RPN) (Figure 7) is calculated by multiplying the SEV, OCC, and DET values together. Multiplication is used in this analysis because they represent a risk tied to impact, likelihood, and reaction. Since each of the three factors can vary from 1 to 10, the product of the values will vary from 1 to 1000. By using multiplication, it is a simple matter to separate the important

RPN – "Risk Priority Number" is calculated by multiplying the SEV x OCC x DET values

Item #	Process name and description	Process Step	Failure Mode	Failure Effect	SEV	Potential Causes	OCC	Verification Method	DET	RPN	Recommended actions
1A-1	Process 1	Step A	Wrong Part	Wasted product, poor or no throughput, low yield	9	Wrong marking	2	Incoming & in-line QC inspection	6	114	
1A-1	Process 1			Wasted product, poor or no throughput, low yield		Unclear drawing or work instruction	5	None	10	450	
1A-1	Process 1			Wasted product, poor or no throughput, low yield		Operator error — training	2	Yearly training sessions,weekly group meetings and briefings	2	36	
1A-1	Process 1			Wasted product, poor or no throughput, low yield		Operator error — too busy	7	None	10	630	

Figure 7 Risk priority number.

from the unimportant while recognizing the importance of all three factors as equal contributors to overall risk to the program. The RPN values can then be sorted from highest to lowest value, providing insight into the most important possible issues. These meet the criteria of highest impact, most likely to occur, and least likely to be detected. This method separates "the vital few" from the "useful many," as the quality pioneer Joseph M. Juran often stated (excerpt from a Public Broadcasting System documentary, "An Immigrant's Gift: The Life of Quality Pioneer Joseph M. Juran" by John Butman and Jane Roessner, Public Broadcasting System).

IMPROVEMENT ACTIONS

Prioritization of Potential Issues

The point of the prioritization is not simply to select from the high-value RPNs; rather, it is to consider the higher-valued items first. It is imperative to review the actions recommended through the lens of business priorities: Are the recommendations actionable? Are they cost-effective? Are there other mitigation factors, such has human safety, to be taken into account? Some of the lower-value RPN items often have simple, low-cost solutions. These should be considered for follow-up as well. These ideas, combined with the prioritization and documented rationale from P-FMECA, provide an excellent basis for making optimal decisions.

First Address Root Cause(s)

Clearly the first place to look for effective improvements is to address the source of the issue—the underlying or root cause. For example, if the cause stems from a lack of tooling to assure proper alignment, then designing the equipment so that special alignment is not required would remove the risk altogether. If this is not possible or too costly, the next step would be to look at alignment tools in the process to decrease the risk of a mistake. "Improved training" is often chosen as a catch-all phrase in corrective actions. That is not to say that training is not important. It is. Certainly, if inadequate training is a factor, it must be addressed. However, the addition of clarifying pictures or photographs and simplified designs or steps, and assuring that the correct tools and fixturing are on hand, are often much more effective actions than general verbiage about signs or slogans, coaching or training.

Inspection is also a useful tool for catching defects, but is often overused as a corrective action. A saying among quality engineers is: You can't inspect quality in! In other words, inspections have limited impact in finding and correcting defects. Corrective actions that prevent defects—that *make it easier to do right and more difficult to do wrong*—are much more effective than added inspection steps. Certainly, such items as improved training and inspections can always be considered as effective additions to selected improvement actions.

Review All Three Factors

1. *Severity:* If a failure were to occur, what could be done to mitigate the impact of the event?
2. *Occurrence:* What can be done to assure that a failure could not happen or at least would be less likely to happen?
3. *Detection:* How can the process be modified to improve the ability to catch a problem in time to prevent it or minimize the consequences?

Present Recommendations to Management

Once the potential improvement recommendations are developed, the next step is to assure that they are well defined in terms that management—program, production, and design management—can understand and use. As with any engineering action items, responsible personnel, scope of effort, and schedule must be defined to assure that the recommendations are handled and reported properly to those making the decisions. These recommendations, then, must be delineated in terms of their estimated costs, so that the decision makers can evaluate the "bang for the buck" in terms of risk.

Follow-Through

As follow-up to this effort, once improvements have been implemented, or to serve as "what-if" studies, further P-FMECA updates may be made, using the initial results as a starting place and then observing the results of reconsidering the risk factors in light of the actions taken.

REPORTING RESULTS

Figure 8 is an example of complete P-FMECA for the simple failure mode and causes displayed in this chapter. Note that the failure mode results have been sorted by RPN value to allow prioritization of the follow-up actions. Once these actions have taken place and their effectiveness has been evaluated, the P-FMECA can then be reviewed for changes to the existing factors to measure RPN improvements.

The completed matrix represents a good review of the scope and level of detail covered, but this is not enough. For the results to be reported properly, a more complete assessment of the ground rules, priorities, and decisions in the performance of P-FMECA serves as a record of the complete effort. The importance of reporting results is manifold:

- The decisions, conclusions, and recommendations developed by the team are communicated in writing.
- The resents reported serve as a record of the underlying rationale behind the possible concerns and improvements developed.
- Results can help track progress after improvement actions have been completed.

Product or product line ____________ Team members ____________

Process or subprocess ____________ Revision date ____________

Item #	Process name and description	Process step	Failure mode	Failure effect	SEV	Potential causes	OCC	Verification method	DET	RPN
1A-1	Process 1	Step A	Wrong part	Wasted product, poor or no throughput, low yield	9	Operator error — too busy	7	None	10	630
1A-1	Process 1	Step A	Wrong part	Wasted product, poor or no throughput, low yield	9	Unclear drawing or work instruction	5	None	10	450
1A-1	Process 1	Step A	Wrong part	Wasted product, poor or no throughput, low yield	9	Wrong marking	2	Incoming & in-line QC inspection	6	108
1A-1	Process 1	Step A	Wrong part	Wasted product, poor or no throughput, low yield	9	Operator error — training	2	Yearly training sessions, weekly group meetings and briefings	2	36

Figure 8 Complete P-FMECA.

- As actions are taken and risks lowered, there will be new "number 1" RPN elements. As follow-up actions are accomplishments, this analysis can serve as a road map for the next priorities and considerations.
- Future P-FMECA procedures can benefit from findings and insights reported from the initial P-FMECA.

SUGGESTIONS FOR ADDITIONAL READING

Electronic Reliability Design Handbook, MIL-HDBK-338B, U.S. Department of Defense, Washington, DC, Oct. 1998.

Ford Motor Company, *Potential Failure Mode and Effects Analysis: An Instruction Manual*, Ford, Detroit, MI, Sept. 1988

Modarres, M., *What Every Engineer Should Know About Reliability and Risk Analysis*, Marcel Dekker, New York, 1993.

O'Connor, P. D. T., *Practical Reliability Engineering*, 3rd ed., J Wiley, New York, 1992.

Procedures for Performing a Failure Mode Effects and Criticality Analysis, MIL-STD-1629A, Notice 3, U.S. Department of Defense, Wahington, DC, Aug. 1998 (canceled).

Chapter 7

FMECA Applied to Software Development

Robert W. Stoddard

INTRODUCTION

Failure modes and effects criticality analysis (FMECA) was developed and used beginning in the 1950s in aerospace engineering, with subsequent adoption and use within the military and nuclear industries, systems safety applications, and reliability engineering applications. An abundance of example uses may now be seen in publications and on the Web related to a myriad of industries, including the medical services and pharmaceutical fields. Essentially, FMECA remains a proven technique for a multidisciplined team to structure thought around anticipating what can go wrong and why, with follow-up thought on how to lessen the chance of occurrence, the severity of the consequence, and the ability of the potential issue to escape detection. FMECA has grown in usage after companies have realized that testing a product is no longer sufficient and that FMECA remains relatively cost-beneficial in light of product liability and recall campaigns [1]. Many companies have adopted FMECA to assess the potential vulnerabilities of both their product design (design FMECA) and their critical business processes or services (process FMECA). Although this chapter delineates how to apply design FMECA to a software product, the literature includes definitions and case studies of the application of process FMECA to the software process, whether it be the software code review process [2] or the software development life cycle in general [2, 3]. Additionally, some recent texts discussing software design for six sigma

Design for Reliability, First Edition. Edited by Dev Raheja, Louis J. Gullo.

(SDFSS) are beginning to scratch the surface in the application of FMECA to software design [4]. In this chapter, a method describing how FMECA may best be applied to software development in the context of different software artifacts and the software development life cycle is discussed.

SCOPING AN FMECA FOR SOFTWARE DEVELOPMENT

FMECA is a tool to structure thought about anticipated failures and their causes. As such, there are a number of ways in which FMECA may be pursued with a software product. Various ways of applying FMECA to software products may best be explained in terms of the software V model shown in Figure 1. In this figure one sees the traditional waterfall life-cycle model of software development beginning with software requirements and passing through software architecture, high-level software design, low-level software design, and code, before beginning the various phases of software testing. This figure delineates the various abstractions of software (requirements through code) which lead to the physical product of source code and an executable image. Although many different software life-cycle models and standards exist, the notion of connecting test activities with their corresponding development activities still applies. Later in this chapter, we discuss the significance to software FMECA in the mapping of test activities to development activities. For now, let's assume that FMECA may be applied at any and all levels of software abstraction. Obviously, the more detailed the abstraction artifacts evaluated, the richer the feedback from the software FMECA. Precedent exists for conceiving of software FMECA within the various abstraction levels of software. In 1990, one author conceived of the application of FMECA to conduct software safety analysis across the different phases of software development [5]. In 2002, an article in the *Encyclopedia of Software Engineering* discussed

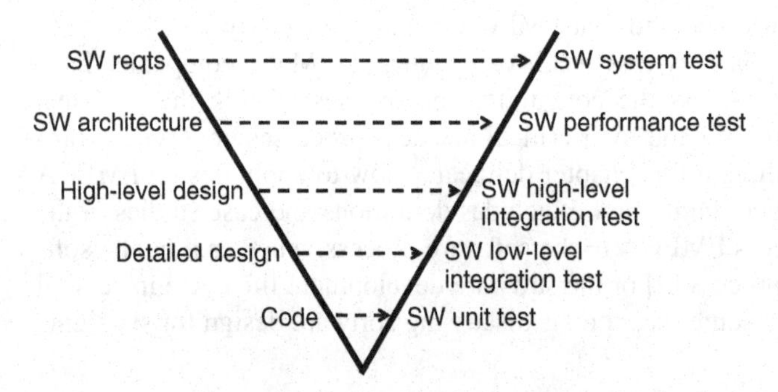

Figure 1 Software V life-cycle model.

using FMECA within two phases of software development: the requirements phase and the development phase [6]. In 2007, another group defined a three-tiered approach to FMECA for electronics systems, including device-, circuit-, and system-level modeling, and used this tiered FMECA approach to refine the prognostic health management for an electronic system [7]. Most recently, in 2010, another group proposed the notion of a "means–ends abstraction" view of any product or system in which decomposition of a product or system may be analyzed using FMECA [8].

The first step in software FMECA involves identifying the business and customer requirements of a software FMECA. The business and customer requirements form the basis for determining the proper scope of the software FMECA. The scope and motivation for software FMECA also depend on the given situation and relationship of the software with the entire product (assuming that the software is embedded in a product). For that reason, a system block diagram or context diagram may be useful in understanding the scoping of the software FMECA. A relationship of a total product- or system-level FMECA with subordinate software and hardware FMECA corresponding to the various components of the product or system is shown in Figure 2. For example, if a top-level product or system FMECA is carried out, the root causes of the failure modes at the product or system level could initiate analysis of the possible failure modes at the next subordinate level. In the same fashion, root causes can lead to itemization of failure modes in a top-down sequence as FMECA is conducted at the subsystem, component, and subcomponent levels. Although depicted as a top-down cascading approach,

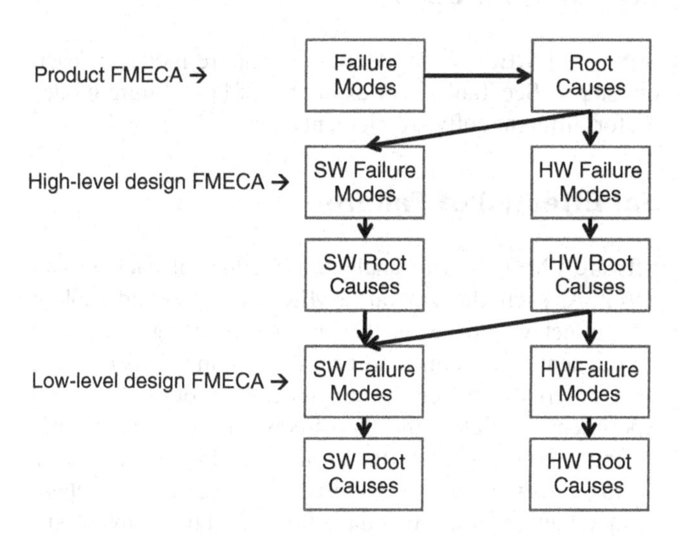

Figure 2 Concept of cascading FMECA.

stand-alone FMECA may be performed at any level of the software abstraction (e.g., the left side of the software V life-cycle model). This may often be the case when a stakeholder desires immediate feedback on the risks to the software without waiting for higher-level FMECA to be conducted. Aligned with this notion, one group in 2007 defined an approach to applying FMECA to software architecture from a perspective of reliability analysis using failure scenarios [9].

FMECA STEPS FOR SOFTWARE DEVELOPMENT

Fifteen steps in software FMECA performed during software development are discussed next.

Step 1: Software Element Identification

As stated earlier, the first step in software FMECA involves identifying the business and customer requirements of software FMECA. The first step focuses on the identification of the software "elements" to be included in the analysis. These elements must be identified in the context of the abstraction level of software FMECA. Table 1 depicts examples of possible software elements to be considered in the FMECA for the various software abstraction levels.

Step 2: Potential Failure Modes

The second step in software FMECA is to identify the failure modes of each software element from step 1. See Table 2 for examples of how failure modes may be discriminated for different software elements.

Step 3: Potential Effect(s) of Failure

The third step in software FMECA is to analyze the effect of each failure mode identified in step 2. As such, the software FMECA team should explore and postulate what the effect would be on the product or system output(s) given that a software element is presumed to be misbehaving, experiencing a failure, or corrupted. Alternatively, one group recently proposed the use of fault injection experiments to determine the effects and severity of specific FMECA causes and failure modes [10]. To achieve this postulation, a multiperspective software FMECA team is required to enable a comprehensive product or system review of the potential effect of the failure mode. The software FMECA team composition must include sufficient perspective

Table 1 Possible Software Elements for FMECA

Software abstraction artifact	Possible software elements
Requirements documentation	Tagged requirements; use cases; features and/or subfeatures; functions; events and transitions between states in state transition diagrams
Architecture documentation	Components from one or more of the seven standard views of software architecture (four views cited as examples): **1.** *Logical view*: modules or objects **2.** *Physical view*: allocation of the software to the hardware processing or communication nodes **3.** *Process view*: concurrency and distribution of functionality **4.** *Development view*: files, libraries, and directories mapped to the software development environment
High-level design documentation	Calls in a software system call tree; McCabe subtrees derived from a software system call tree; predominant interactions between major software objects in an object-oriented development; major threads of execution in real time, interrupt-driven embedded software; relations in object relationship diagrams or entity relationship diagrams; message sequences in message sequence charts; specific action times in timing diagrams
Low-level design documentation	Low-level calls among detailed software components, subroutines, libraries; interfaces between low-level software components; control and data flows between low-level software components
Source code (e.g., programmed language)	Variables; constants; pointers; arrays; other data storage or memory items; key decisions in code; key calculations in code; key inputs or outputs of a given software component

to accommodate this postulation, depending on the original scope of the software FMECA. If the scope includes the complete product or system operation, the software FMECA team must include participants who understand the operation of the software *and* the software's role in the overall product or system operation. In practice, the scope will be either limited to the software itself or to the entire product or system. Obviously, the scope significantly influences what the effect and subsequent severity rating will be. Refer to Table 3 for examples of the potential effects for some failure modes.

Table 2 Failure Modes for Software Elements

Software element	Potential ways to think of failure modes
Tagged requirements; use cases; features and/or subfeatures; functions	The visible deficient outcomes possible with a given requirement, use case, feature, subfeature, or function
Components from one or more of the seven standard views of software architecture	The visible deficient outcomes possible with a given component
Calls in a software system call tree	The visible deficient response to a call in a software system call tree
McCabe subtrees derived from a software system call tree	The visible deficient outcome or effect of a given McCabe subtree operation
Predominant interactions between major software objects in an object-oriented development	The visible deficient outcome or evidence of a deficient interaction between major software objects
Major threads of execution in real-time, interrupt-driven embedded software	The visible deficient outcome or side effect of a major thread of execution
Low-level calls among detailed software components, subroutines, or libraries	The visible deficient outcome or side effect of a low-level call
Interfaces between low-level software components	The visible deficient outcome or side effect of the exercise of an interface between low-level software components
Variables; constants; pointers; arrays; other data storage or memory items	The visible deficient outcome, consequence, or side effect of a flawed or corrupted variable, constant, pointer, array, etc.
Key decisions in code; key calculations in code	The visible deficient outcome, consequence, or side effect of a flawed decision or calculation performed in the code
Key inputs or outputs of a given software component	The visible deficient outcome, consequence, or side effect of a flawed or corrupted key input or output of a given software component

Step 4: Severity Rating

The fourth step in software FMECA is to specify the criticality or severity for each failure mode in step 3. The criticality or severity measure is normally represented on a scale of 1 to 10, with higher values representing greater criticality or severity. One source in 2002 defined a comprehensive methodology for architecture-level reliability risk analysis that promoted the use of

Table 3 Effects of Potential Failure Modes for Software Elements

Software failure mode	Effects of the corresponding software failure mode
Use case: User unable to update coordinates for firing position.	Weapon will be firing at the wrong target and miss the intended target.
Software architecture physical view component: The processor handling all sensor data and calculations experiences a failure, thereby depriving the flight navigation processor software of real-time updates to sensor data.	Without real-time sensor data updates, the navigation processor software may worsen the stall situation rather than correct the stall situation, enhancing the chances of a crash.
Software system call tree: A software system call to the mission planning subsystem experiences a lack of response to the call due to a mission planning subsystem deficiency.	The mission planning subsystem does not conduct an update to react to the battlefield situation, thereby endangering the life of the aircraft.
McCabe Subtree: One of the unique, logical McCabe subtree executions within a software call tree misbehaves and does not update a redundant airspeed sensor reading.	The redundant airspeed sensor reading is flawed but will only affect the aircraft operation in the event that the primary airspeed sensor fails.
Interaction between two software objects: Two objects exist as examples: an event handler and a sensor communicator. The sensor communicator sends an event to the event handler. However, the event handler, for some reason, ignores the event until the end of an ongoing, lengthy, low-priority task.	The effect of the mishandled event by the event handler may be serious or minor, depending on the real-time constraints on the software system. In a collision-avoidance situation, the effect could prove catastrophic!
Pointers: Within a unit of code responsible for the weapons targeting mechanism, a pointer is somehow misdirected and points to an incorrect part of memory, which then provides a false value needed for targeting.	The false value could cause the weapon to misdirect and miss its target, *or* it could actually confuse friend and foe, with disastrous results.

a heuristic risk assessment methodology based on a wide variety of dynamic software architecture complexity metrics to establish severity ratings [11]. From the software FMECA standpoint, there is little difference here from a traditional hardware FMECA, but with a reminder that the scope of software FMECA will heavily influence the effect and this severity rating. For many reasons, it remains prudent to adopt a severity scheme that is consistent with the severity scheme used in hardware and product or system FMECA.

Table 4 Causes of Potential Failure Modes

Failure mode	Potential root or underlying cause(s)
Use case: Weapon will be firing at the wrong target and miss the intended target.	Requirements ambiguity exists about when a user is able to change the coordinates for the firing position.
Software architecture physical view component: Without real-time sensor data updates, the navigation processor software may worsen rather than correct the stall situation, enhancing the chances of a crash.	The sensor processor experiences an intermittent failure due to power surge in the hardware electronics.
Software system call tree: The mission planning subsystem does not conduct an update to react to the battlefield situation, thereby endangering the life of the aircraft.	The mission planning subsystem fails to respond to the system call because it is programmed to react in minutes rather than seconds (e.g., two different scales of time and priority are programmed in the system).
McCabe subtree: The redundant airspeed sensor reading is flawed but will affect aircraft operation only in the event that the primary airspeed sensor fails.	Flawed decision logic within the McCabe subtree causes an outdated airspeed sensor reading to be sent as the redundant airspeed reading.
Interaction between two software objects: The effect of the mishandled event by the event handler may be serious or minor, depending on the real-time constraints on the software system. In a collision-avoidance situation, the effect could prove catastrophic!	The programming logic in the event handler mistakenly assigns a very low priority to the event received from the sensor communicator.
Pointers: The false value could cause the weapon to misdirect and miss its target, *or* it could actually confuse friend and foe, with disastrous results.	A unit of code contains faulty conditional logic that resets the pointer to an invalid area of memory.

Step 5: Potential Cause(s) of Failure Modes

The fifth step in software FMECA is to identify the potential root or underlying cause(s) of each failure mode identified previously. The nature of the potential root or underlying cause(s) varies greatly by the nature of the failure mode and the level of abstraction of software FMECA. Table 4 depicts some examples of causes of each failure mode described in Table 3.

Step 6: Failure Occurrence Rating

The sixth step in software FMECA is to establish the probability of occurrence of the root or underlying cause identified previously. Within traditional FMECA operation, the FMECA team usually assesses historical failure data to assess the probability or likelihood of occurrence of the cause. Within software FMECA, the team may use a number of different ways to establish the occurrence rating. To begin with, a combination of historical test and/or field failure data may be used to establish the rating objectively. However, for most software FMECA, objective occurrence data usually does not exist, especially at the different levels of abstraction. This was noted in 2002 by one author as a major impediment to conducting FMECA for software [6]. However, in these cases, we now propose that the software FMECA team use subjective expert opinion or surrogates for the occurrence ratings. For example, a team establishing occurrence ratings at the code level of abstraction for a software FMECA may use the Keene model, based on capability maturity model assessment results, as a valuable surrogate for assessing the failure potential of individual software units prior to accumulation of actual empirical data from test environments or customer application environments. The Keene model is described in detail in Chapter 5. As an alternative to the Keene model, McCabe cyclomatic complexity may be useful to assess the occurrence rating. The McCabe method is based on mathematical graph theory and quantifies the number of unique logical paths in a software unit. This measure has been used as a risk measure by many defense and industry organizations to plan and evaluate software unit testing coverage. After the software FMECA team identifies the occurrence rating for the software units associated with the causes in step 5, the team may then identify a software unit with a known field failure experience, or an expert consensus of an occurrence rating, to be the benchmark. However, a word of caution deserves attention by software FMECA teams utilizing surrogate measures to establish the occurrence rating. The surrogate values are meant to be applied early in the development process and should be updated with higher-confidence data later in the development process. The value of the risk priorit number (RPN) rating system is not in the absolute ratings assessed but, rather, in the relative ratings assessed. The RPN scores are meant to distinguish high from medium from low risk so that management can take action on the priority high-value RPN values. As such, a surrogate measure should provide a reasonable differentiation of occurrence scores. The surrogate measure adds little value to the RPN exercise if all of the software units receive essentially the same occurrence scores. The reader should examine Table 5 for additional ideas on possible surrogate values for occurrence ratings at different software abstraction levels.

Table 5 Possible Surrogate Values for Software Occurrence Ratings

Abstraction level	Possible ideas for surrogate measures of occurrence
Requirements documentation	Ambiguity scores attached to individual requirements or use cases; reading level and comprehension scores attached to individual requirements or use case descriptions; historical requirements volatility measures from similar products for corresponding requirements and use cases
Architecture documentation	Components from one or more of the seven standard views of software architecture (*logical view:* complexity measures of modules or objects; *physical view:* relative size or complexity of the hardware processing or communication nodes; *process view:* subjective assessment of the expected degree of usage of functionality; and *development view:* volume of development and/or maintenance changes to individual files, libraries, and directories that are mapped to the software development environment)
High-level design documentation	Dynamic execution reachability metrics; Keene model; McCabe design complexity (S0); McCabe integration complexity (S1)
Low-level design documentation	Dynamic execution reachability metrics; Keene model; McCabe module design complexity
Source code (e.g., programmed language)	Keene model; McCabe cyclomatic complexity; Halstead complexity; data flow complexity

Step 7: Detection Rating

The seventh step in software FMECA is to assess the probability that a given root or underlying cause would escape detection and ultimately affect an output (based on the scope of software FMECA). A number of development and test activities, as well as design and code conventions, are most critical to preventing defective or misbehaving software "elements" from affecting output(s) of the software or product or system (depending on the scope of software FMECA). Table 6 describes some common key activities and conventions associated with the different software abstraction levels.

In traditional FMECA operation, the FMECA team would normally access historical data related to the detection of root or underlying causes and base the detection rating on those hard data. However, most software FMECA

Table 6 Software Detection Activities and Methods

Abstraction level	Detection activities and conventions
Requirements documentation	Software requirements peer reviews; software requirements ambiguity reviews; formal software requirements reviews with stakeholders; software system test; customer and/or user test
Architecture documentation	Software architecture peer reviews; software architecture trade-off analysis method; formal software architecture reviews with stakeholders; software performance test including test of software quality attributes and scenarios; software performance modeling and simulation; strategic software fault tolerance and fault recovery mechanisms
High-level design documentation	High-level software design peer reviews; formal high-level software design reviews with stakeholders; high-level software integration test; high-level software design fault tolerance and fault recovery mechanisms
Low-level design documentation	Low-level software design peer reviews; low-level software design reviews with stakeholders; low-level software integration test; low-level software design fault tolerance and fault recovery mechanisms
Source code (e.g., programmed language)	Code peer reviews; software unit test; defensive programming practices unique to each computer language

teams find themselves with little, if any, software data of that nature. As a result, software FMECA teams have found it useful to follow a structured approach: assessing the detection rating subjectively using a quantitative scheme leveraged from hardware reliability block diagrams. The software team first delineates the complete list of applicable detection activities and conventions. This list should encompass all the abstraction levels applicable to the organization's software development, test, and maintenance life-cycle activities and practices as well as conventions with regard to fault tolerance and recovery.

Next, for each individual root or underlying cause, the software FMECA team then subjectively assesses the independent probability that each detection activity or convention will detect or stop the root or underlying cause from affecting an output (in terms of the scope of software FMECA).

Finally, the software FMECA team calculates the detection score for each underlying cause by treating the list of applicable detection activities and conventions as a parallel reliability block diagram. In this manner, the overall probability of an underlying cause escaping complete detection is computed as follows:

$$\text{probability of overall escape} = [1 - p(1)][1 - p(2)][1 - p(3)] \cdots [1 - p(n)] \quad (1)$$

where n is the number of detection activities and conventions, $i = 1 \cdots n$, and $p(i)$ is the probability of the ith activity or convention detecting the underlying cause. With this computation, the probability of overall escaping detection for each root or underlying cause may easily be mapped to the detection rating of 1 to 10. However, as stated earlier, the purpose of the RPN scoring is to differentiate RPN scores so that management may take action on the highest-priority risk items. With that said, a straight mapping of the overall probability of detection to the detection rating of 1 to 10 may not be recommended. An alternative mapping scheme may be needed so that differentiation is evident. Because of this aspect, it would thus be invalid and misleading to compare software FMECA RPN scores to scores from a hardware (mechanical or electrical) FMECA. In practice, the approach described above for software FMECA teams seems to work effectively and expediently. Software FMECA team members were able with amazing speed to proceed to assess detection ratings for each underlying cause.

Step 8: Calculating the Risk Priority Number

The eighth step in software FMECA is to calculate the RPN. This is a standard convention within the FMECA process in which the three scores (severity, occurrence rating, and detection rating) are multiplied together to get an overall risk score (the RPN). Software FMECA teams most often implement the standard industry convention of the value range 1 to 10 for each score, which then produce RPN scores in the range 1 to 1000. The FMECA method usually then requires an FMECA team to establish an RPN threshold such that any RPN score above the threshold requires additional treatment, as discussed in subsequent steps involving prevention and mitigation actions. The industry norm for an RPN threshold using a value range of 1 to 10 on each individual score is 200. However, the software FMECA team may decide to select a lower threshold to identify a larger portion of items requiring prevention and mitigation action planning in the upcoming FMECA steps.

Step 9: Improvements Related to Mitigation

The ninth step in software FMECA involves a critical software FMECA team brainstorming on how to mitigate or lessen the effects of each cause and/or failure mode. Depending on the nature of the cause and failure mode, a variety of possible mitigation actions may be taken. Generally, the team will identify ways in which the software and/or product or system may be designed to be more robust and insensitive to the software cause and/or failure mode. Standard thinking surrounding design margins and redundant mechanisms applies equally in the software domain, thereby reducing the severity of the underlying cause.

Step 10: Updated Severity Rating

The tenth step in software FMECA involves reassessing the severity rating by assuming that the mitigation actions just identified will be implemented successfully. The updated severity rating now depicts the reduced risk resulting from the mitigation actions.

Step 11: Improvements Related to Prevention

The eleventh step in software FMECA involves a critical software FMECA team brainstorming on how to prevent underlying causes from occurring. Preventive actions may include process and training changes related to software development, test, and maintenance activities. Preventive actions may also include software design and code implementation that identifies and stops causes immediately at the point of origin. There may be a fuzzy line of demarcation between prevention improvement and the upcoming discussion of detection improvement. I choose to include the immediate capture of a cause as prevention, whereas subsequent activity falls in the realm of detection. In any case, software FMECA team members should make use of Poke-Yoke and other mistake-proofing methods to creatively identify prevention actions for the various causes.

Step 12: Updated Occurrence Rating

The twelfth step in software FMECA involves reassessing the occurrence rating by assuming that the prevention actions just identified will be implemented successfully. The updated occurrence rating now depicts the reduced risk resulting from the preventive actions.

Step 13: Improvements Related to Detection

The thirteenth step in software FMECA involves critical software FMECA team brainstorming on how to detect and stop the hypothetical underlying causes from escaping through the development, testing, and maintenance process, and subsequent operation of the software, to affect an output of concern (based on the scope of the software FMECA). At this point, the Software FMECA team will need help from software process experts to identify what new or modified software processes would detect the causes earlier and more assuredly. Additionally, software architects and designers will need to advise the software FMECA team on what new or modified software architecture components, design structure, or code is needed to detect causes during software runtime. Often, the recommendations from this step form the needed improvement plan for enhanced fault tolerance and recovery, including the classes of exceptions and the exception handling expected.

Step 14: Updated Detection Rating

The fourteenth step in software FMECA involves reassessing the detection rating by assuming that the detection improvement actions just identified will be implemented successfully. The updated detection rating now depicts the reduced risk resulting from the detection improvement actions.

Step 15: Updated RPN Calculation

The fifteenth step in software FMECA is to update the RPN numbers calculated for each of the FMECA line items that were above the established RPN threshold, and for which additional preventive and mitigation actions were identified. For this step, the software FMECA team reviews all of the preventive, mitigation, and detection actions recommended for a given FMECA line item and then computes the updated RPN value based on the three updated ratings. The updated RPN score now reflects the overall revised risk assessment. In most organizations, the initial and revised RPN numbers form the risk profiles that management finds convenient to manage during the development and fielding of products. Often, management will monitor the volume and age of outstanding RPN values above the acceptable threshold so that they can ensure that proper engineering resources are applied to preventive, mitigation, and detection activities and conventions. A number of organizations also use the results of the initial and updated RPN scores to feed more advanced models that seek to predict project and product success outcomes, such as schedule and cost performance, customer satisfaction, and even early product returns.

IMPORTANT NOTES ON ROLES AND RESPONSIBILITIES WITH SOFTWARE FMECA

A number of points are noteworthy related to roles and responsibilities when conducting successful software FMECA. First, a facilitator other than the author of the software artifact is needed to ensure that the proper software FMECA process is followed in a timely fashion with real-time resolution of team conflicts. The primary reason for the independent facilitator is a lack of bias toward the software artifact under review. Generally, authors of software artifacts are success-oriented and do not find it comfortable to lead in-depth discussions of the potential issues with their own artifacts. The best facilitator frame of mind is one of an unrelenting detective, prompting the team to look deeper and deeper, questioning all assumptions made by the author.

Second, the software FMECA team members must possess domain knowledge of the software artifact under review. Without domain knowledge they will add little to what the independent facilitator contributes.

Third, the immediate supervisor and management chain of the author of the software artifact under review must abstain from participation in software FMECA. In this fashion, open and candid discussion may occur about potential issues in the software artifact without the worry of recrimination.

Fourth, a test perspective must be represented on the software FMECA team to ensure that a critical test and evaluation mindset is available to help probe for potential issues in the software artifact. Again, good test engineers have a critical, detection-oriented mindset energized to find any weakness, flaw, or vulnerability. This is necessary, as software developers authoring code tend to have sunny day–biased views of their software.

LESSONS LEARNED FROM CONDUCTING SOFTWARE FMECA

A number of lessons have been learned from conducting software FMECA with clients. Although some have been touched upon earlier in the chapter, the full list is discussed below for clarity and reinforcement.

1. Software FMECA teams have found that the FMECA process works most easily when the team populates the software FMECA column by column. The rationale is that the team benefites by at least getting all of the software elements identified before jumping into the failure modes. Essentially, this reduces the amount of context switching needed to populate the software FMECA template. At times the columns would be conducted as pairs in the following order:

a. Software elements
b. Failure modes
c. Effects of failure modes in tandem with the discussion of the severity rating
d. Causes in tandem with occurrence ratings
e. Detection rating
f. Initial RPNs calculated

g. Prevention improvements
h. Mitigation improvements
i. Detection improvements
j. Final RPNs calculated

2. When briefing the results of software FMECA, teams found it easier to brief the audience in row fashion rather than in column fashion, to minimize context switching during the debrief.

3. Software test personnel seemed to have a better background and perspective to facilitate software FMECA workshops. Their critical detective and persevering thinking skills, along with their perspective of test to failure, promoted more thorough software FMECA sessions.

4. The results of software FMECA proved invaluable to the software test community, corroborating what other authors had seen previously [6]. They often rejoiced at the list of vulnerabilities from software FMECA because the list became a great targeting mechanism to drive resource- and schedule-constrained software testing.

5. The results of software FMECA served as rich input to even the most basic thinking of error-handling and fault-tolerant design practices, which also corroborated what other authors discovered in 2005 [12]. Even the most resistant software developers of defensive design and coding found it compelling to at least defend against the highly rated items from the software FMECA workshop.

6. Software FMECA is indeed quite different from traditional software peer reviews and inspections. Table 7 summarizes the most notable differences evident during the client pilots of software FMECA.

7. The authors of the software artifacts studied by a software FMECA team must exhibit much patience. Whether the authors are software requirements analysts, software system engineers, software architects, software

Table 7 Software Inspections and FMECAs Contrasted

Peer reviews and formal inspections	FMECA
Require and depend on significant advance individual preparation	Requires less advance preparation by participants
Focus on established standards and conventions from a compliance standpoint (closed focus)	Focuses on brainstormed failure modes (open focus)
Focus exclusurely on identifying noncompliance (defects)	Focuses on failure modes and corresponding root causes
Look only at actual issues or defects (i.e., they have already occurred)	Purposely anticipates issues and defects that have yet to occur

designers, or programmers, they often find it uncomfortable to participate in a detailed critique of their work focused predominantly on a hypothetical critique of potential weaknesses. They are success oriented and often believe that software FMECA will be a waste of time. However, experience generally shows that if they are patient and endure the workshop, they will invariably be grateful for finding the one golden nugget of risk that they had overlooked prior to the workshop. Many software developers reported that they were indeed thankful for finding the risk early before the risk is realized and becomes a headline in product returns. Indeed, my personal experience is that almost all software FMECAs will find at least one significant potential risk to be addressed. This is in sad contrast to my firsthand witnessing of hundreds of software inspections that failed to uncover any significant issues. For this reason, within the development community, software FMECA generally does gain favor over traditional software inspection because it helps software developers convincingly.

8. As the reader undoubtedly gathers by now, the proper scoping of software FMECA is paramount. In addition, it is also paramount to establish whether there will be a series of FMECA analysis, possibly in the cascading fashion described earlier in the chapter. Deciding on these early saves enormous time and effort and enables FMECA to be accomplished in the best order and maximize designer inputs for a given FMECA analysis.

9. FMECA in general, and especially software FMECA, served to bolster a proactive, pragmatic, institutional learning mechanism within organizations and across programs, projects, and product development teams. Several clients decided to create a set of software FMECA templates with a unique template for each level of the software abstraction. In this fashion, a unique software FMECA template with pre-populated software elements, failure modes, and underlying causes was used during each major phase of the software development life cycle. During each software development life-cycle phase, the software team then self-assessed the three ratings (i.e., occurrence, severity, and detection) of the prepopulated items in context of their current software. The result was that the software team learned from earlier software teams' experiences of failure modes and underlying causes. Experience with several clients demonstrated that this approach enabled a living lessons-learned approach for product development teams and ended the seemingly persistent common failure modes reappearing across generations of similar products.

CONCLUSIONS

Software FMECA may be applied in a number of flexible ways at different points in software development and maintenance life cycles. Although still novel to the software development community in general, software FMECA is steadily gaining acceptance with development teams as a superior method of improving software quality. Additionally, software FMECA has served to bridge the cultural and process barriers that separate software developers and software test staff. Indeed, one may see similarities of the unified involvement of developers and testers in software FMECA with the strengths of agile, paired, and extreme programming. Hopefully, both software and hardware professionals will come to see the potential of software FMECA and will encourage their software organizations to adopt software FMECA.

REFERENCES

[1] Auguston, K., What went wrong at Toyota? *Des. News*, vol. 65, no. 3, 2010, pp. 31–33.

[2] Dillibabu, R., and Krishnaiah, K., Application of failure mode and effects analysis to software code reviews: a case study, *Software Qual. Prof*., vol. 8, no. 2, Mar. 2006, pp. 30–41. Retrieved Nov. 1, 2011, from ProQuest Computing (Document ID: 1006414181).

[3] Laprie, J., and Littlewood, B., Viewpoint. *Commun. ACM*, vol. 35, no. 2, 1992, pp. 13–21.

[4] Research and Markets: Software design for Six Sigma: a roadmap for excellence, *Business Wire*, Dec. 8, Retrieved Nov. 1, 2011, from ProQuest Newsstand (Document ID: 2207815221).

[5] Hansen, M. D., Software: the new frontier in safety, *Prof. Safety*, vol. 35, no. 10, 1990, p. 20. Retrieved Nov. 1, 2011, from ABI/INFORM Global (Document ID: 721529).

[6] Chidung, L., Failure modes and effects analysis, in J. J. Marciniak, Ed., *Encyclopedia Of Software Engineering*, Wiley, Hoboken, NJ, 2002. vol. 1, pp. 1521–525.

[7] Brown, D. W., Kalgren, P. W., Byington, C. S., and Roemer, M. J., Electronic prognostics: a case study using global positioning system (GPS), *Microelectron. Reliab*., vol. 47, no. 12, 2007, pp. 1874–1881.

[8] Lee, J., Katta, V., Jee, E., and Raspotnig, C., Means–ends and whole-part traceability analysis of safety requirements, *J. Sys. Software*, vol. 83, no. 9, 2010, pp. 1612–1621.

[9] Tekinerdogan, B., Sozer, H., and Aksit, M., Software architecture reliability analysis using failure scenarios, *J. Syst. Software*, vol. 81, no. 4, 2008, pp. 558–575.

[10] Grunske, L., Winter, K., Yatapanage, N., Zafar, S., and Lindsay, P. A., Experience with fault injection experiments for FMEA, *Software Pract. Exper*., vol. 41, no. 11, 2011, pp. 1233–1258.

[11] Yacoub, S. M., and Ammar, H. H., A methodology for architecture-level reliability risk analysis. *IEEE Trans. Software Eng*., vol. 28, no. 6, 2002, pp. 529–547.

[12] Hammarberg, J., and Nadjm-Tehrani, S., Formal verification of fault tolerance in safety-critical reconfigurable modules, *Int. J. Software Tools Technol. Transfer*, vol. 7, no. 3, 2005, pp. 268–279.

Chapter 8

Six Sigma Approach to Requirements Development

Samuel Keene

EARLY EXPERIENCES WITH DESIGN OF EXPERIMENTS

The present author was fortunate to have a positive defining reliability experience in the 1970s related to design of experiments (DOE) as the foundation for six sigma and design for six sigma (DFSS) processes and tools. I was involved with a team designing one of the first laser scanners. The active element was a HeNe laser, the type with a red beam that we see regularly in foodstore checkout lanes. Early HeNe lasers were chaotic in their parametric performance and their demonstrated lifetimes. Up to that time, these lasers had been used only under laboratory conditions and at construction sites; there had been no emphasis on high-reliability lasers. The industrial application of a photoscanner demanded higher reliability.

High reliability could only be achieved with testing and design changes, which would prove to be a great burden to the program. The test samples had a long delivery time and were expensive, about $5000 apiece. Their performance, sample to sample, varied widely. There were differences in their starting voltage, run current, and important from a reliability standpoint, in their lifetimes. These qualities are labeled *key process output variables* (KPOVs) in the six sigma vernacular. The life goal was 5000 hours before

Design for Reliability, First Edition. Edited by Dev Raheja, Louis J. Gullo.

replacement. The laser test samples that were exercised in initial tests were exhibiting 50 hours on average, with maximum lifetimes between 100 and 1000 hours across 50 samples. Further testing was conducted on new samples. When a laser sample was finally observed exhibiting stable characteristics across several thousand hours of operation, it represented proof that the lifetime and desired stability were obtainable. It proved the existence theorem: that lasers could achieve the life goal. All the laser failures were analyzed thoroughly and some subtle differences were found. Some lasers had 2024 aluminum cathodes; some had 6067 aluminum cathodes. Some were temper T4; some were temper T6. These lasers were built differently even though the purchase specifications and the vendor's datasheet did not identify or allow for such variations. These were just variations that the supplier did not feel were important and had not been a factor to his customers up until that time. The vendor was not even aware of the subtle changes in his supplier's materials. These input variables are potentially *key process input variables* (KPIVs). We desire to understand the effect of these input variables on the output(s) of the laser and to determine what the key input process variables are.

Teaming with the supplier, metallurgists, and failure analysis and development personnel, we identified the possible variations in process and materials in their product. These cross-functional meetings even identified other configuration changes that the manufacturer was considering. Twenty-five potential KPIVs were identified. This list was scaled down to 13 potentially more significant variables that could influence laser performance. These process and product variations were all analyzed in a series of designed experiments. The first experiment was a screening experiment to find the more important KPIVs. This test simultaneously examined 10 possible input factors at two levels (e.g., 6061, 2024 aluminum) each and three other factors at three levels (e.g., high, medium, and low fill pressures). These variables were tested in the screening test at their extreme levels to see if there was any impact on laser performance. The effects of these input variables were measured on three critical output variables: laser start voltage, laser current stability, and laser life. This initial screening test narrowed the initial 13 KPIVs from the screening test down to a set of six KPIVs. Subsequent testing then established, and later validated, the optimum setting of each of these KPIVs to achieve the best KPOVs. Through the systematic effort of DOE with design requirement enhancements and manufacturing process improvements, the laser scanner performance over an entire population of lasers in its application went from chaotic to rock solid. The new laser scanners never caused a field problem.

Conversely, too often an operating point is set based on a single sample or single lot of samples from the supplier. Supplier part variation occurs later, compromising product performance. Redesign too often compromises the original design architecture and increases design complexity. DOEs are surprisingly underutilized. One might ask why a successful tool such as DOE is not embraced more widely. It could be that developers don't use this tool for a number of reasons. It may be that developers and their management:

1. Don't know of DOE.
2. Don't have the statistical skill sets or confidence to use it.
3. Don't want to expend the time and effort to design and conduct a formal experiment.
4. Believe that the problem is trivial and that only one variable is involved (i.e., it is too simple for a sophisticated tool such as DOE). They want to treat the problem as a one-factor-at-a-time (OFAT) problem.
5. Feel that their intuitive solution is adequate.

Yet in the end, DOE is the *only* way to analyze a multifactor problem with statistical confidence and reveal the factor interactions. The factor interactions are never uncovered in an OFAT problem, and they are often the predominant drivers of the effects measured. An interaction is prevalent whenever some effect is conditional. For example, a question that illustrates interactions is: "Do you feel better with humidity added to your environment?" The answer is: "It depends." There is an interaction, but how and when? People living in their houses will add humidity in the wintertime when the house is heated to 66 to 70°F. Humidity is undesirable under extreme cold or extreme heat conditions. An engineering example of factor interaction is the problem experienced in underinflated Firestone tires on a Ford Explorer as related in *Forbes* magazine on June 20, 2006 (see http://www.forbes.com/2001/06/20/tireindex.html). This led to tire pressure blowouts whenever *all* three conditions were present: (1) underinflated tires, (2) Firestone tires, and (3) Ford Explorer automobile. This is an interactive problem and led to Firestone tire recalls in 2000.

I call DOE the sweet spot of six sigma! Through DOE the design variation can be explored and a proper, stable design point established as was done for the HeNe laser build. This became a collaboration exercise, with the supplier's designers and manufacturing people meeting with the system design and material research staff from our company. There was a real-time learning experience for all participants. The device supplier better understood the system application needs for their device. The system folks learned the build constraints on the laser as well as potential future design and material changes under consideration. The research staff reported the changes and variation they found in the laser samples, some of which were unknown to the laser supplier. So the DOE not only addressed the immediate design and material variations in the laser construction but provided a platform from which to investigate future changes. DOE helped establish the optimum laser build design point by properly defining the materials and processes used to build it. This process yielded a product design that met all of its functional and reliability goals and provided an understanding of how to control potential product variations. This has been the six sigma approach. Six sigma improves the product as well as its underlying development process. This has been the most satisfying design experience in my career.

SIX SIGMA FOUNDATIONS

Six sigma originated at Motorola in the early 1980s in response to a CEO-driven challenge to achieve a 10-fold reduction in product-failure levels in five years. Meeting this challenge required swift and accurate failure analysis and correction of the underlying cause. In the mid-1990s, Motorola divulged the details of their quality improvement framework, which has since been adopted and widely embraced by most large manufacturing companies. The process improvement is aided with six sigma, setting management's expectations for developing better products, harvesting opportunities, and being better able to resolve chronic problems. The six sigma legacy increases management and engineering receptivity to opening doors for improving processes and products. Six sigma emphasizes doing the right thing and doing things right. Its goal is to improve the product and the underlying process that developed the product. So it fixes the current problem and sets the stage for future products to benefit by avoiding the cost associated with solution of the same systemic problem.

The Meaning of Six Sigma

Statistically, *sigma* represents the standard normal variate or standard deviation. The sigma count indicates, in terms of the number of standard deviations, how far away the center of the distribution is from the specification limit. Figure 1 illustrates six sigma (6σ) process capability, which means that the constraining process limit is separated from the mean by 6σ. A normal distribution contains practically all of its data points within $\pm 3\sigma$. To be exact, $\pm 6\sigma$ about the mean contains 99.72% of all the population data points, so 0.28% of all points lie outside the $\pm 3\sigma$ distribution limits. The distribution asymptotically approaches zero in the limit but never gets there. Any points falling outside the specification limits are deemed to be unacceptable and are labeled as failures. When the constraining limit [i.e., the upper spec limit (USL)], shown in Figures 1 and 2, is 6 standard deviations from the mean, the number of failures is on the order of a few parts per million. The 3σ and 6σ capability processes are illustrated in Figures 1 and 2. Compare the safeguard of this process with that of a 6σ process. Notice that the distribution tail is well within the USL, with only 3.4 ppm outside the specification limit.

A 6σ process will exhibit a short-term defect rate of approximately 1 in a billion opportunities. There are shifts and drifts of the process over time. This has typically been observed to amount to a 1.5σ shift over time. So a 6σ process

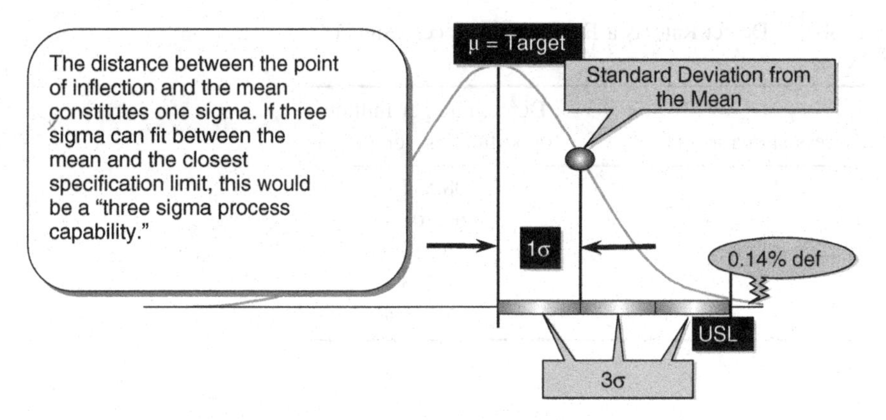

Figure 1 A 3σ process.

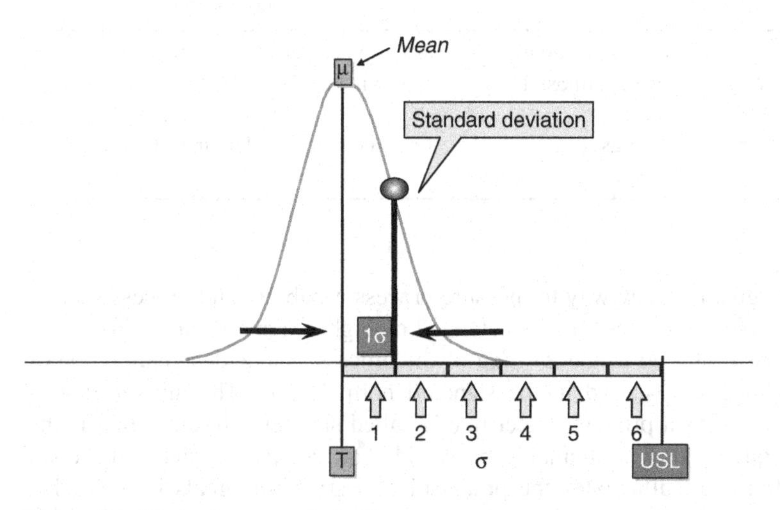

Figure 2 What a 6σ process looks like.

in the short term really equates to a 4.5σ process long term when accounting for shifts in the process. This long-term average defect rate amounts to 3.4 defects per million opportunities. This amount of shift and drift is considered nominal unless better data indicate otherwise. There are a few existing 6σ processes. Several Carnegie Mellon University Software Engineering Institute Capability Maturity Model Integrated (CMMI, a registered trademark) level 5–rated companies assert that their software code can get that good, over time, as the code is refined. Certainly, airplanes have engineered their products to exceed 6σ. The importance of the sigma level on the process defect rate is illustrated in Table 1.

Table 1 Defect Rate as a Function of Process Sigma Level

Process sigma level	Defect rate per million opportunities, long term
2	308,537
3	66,807
4	6,210
5	233
6	3.4

Table 2 Practical Implications of Process Sigma

99% good (3.8σ)	99.99966% good (6σ)
20,000 lost pieces of mail per hour	7 lost pieces of mail per hour
Unsafe drinking water for almost 15 minutes a day	Water is unsafe 1 minute every 7 months
Two short or long landings at most major airports each day	One short or long landing every 5 years

Six sigma is a new way to measure process capability and process robustness. The best processes are characterized by higher sigma values. This sigma value is also referred to the *Z-value* of the process. The Z designation comes from its measure as used in the standard normal table. This universal way of benchmarking a product defect rate is called six sigma because that is the level of quality that is attainable by world-class companies, and thus sets a standard. Table 2 illustrates the practicality of six sigma goals in areas that are critical to our lives and lifestyles.

THE SIX SIGMA THREE-PRONGED INITIATIVE

Six sigma is a mindset of a commitment to use data-driven approaches to problem definition and resolution. It is typically thought of as a statistical method. It is that, but it is more. Six sigma is made up of processes to map tools systematically to define the problem (opportunity) and then move forward to resolve the problem or harvest the opportunity. *Design for six sigma* (DFSS) is the term applied to building world-class designs using six sigma tools. DFSS is the major focus of this chapter. Suffice it to say for now that six sigma tools can be categorized into three groups.

1. *Collaborative tools*. Some key tools here: mind maps, swim lane diagrams, flowcharts, and SIPOC (suppliers, inputs, process, outputs, and customers), fishbone, and control diagrams. These tools are strongly visual and most helpful in developing a fullness of requirements. Collaborative tools also promote cross-functional communications and leave a legacy and traceability to the design process. They provide completeness to requirements development. These planning exercises are close ended. Nonstructured planning exercises can go on endlessly. These collaborative tools have a known completion. They also establish requirements planning artifacts for traceability and future reference. They are monuments to the planning process. Collaborative tools support good product planning by aiding:

- **a.** *Situational awareness*: of customer needs and the environmental influences on a product (external to the product)
- **b.** *Navigability*: through the design components and interfaces (internal consistency, interfaces)
- **c.** *Product understandability*: to reduce interoperability problems and improve robustness to design changes. This also helps to assess the completeness of the design and identify improvement opportunities.
- **d.** *Compact focus*: to see the overall design in a single view or a more limited view (if more frames are required)
- **e.** *Differing system views*: to get "fresh" examinations of the system under development, looking for improvement opportunities and design completeness

2. *Analytical or quantitative tools*. Some key tools here: failure modes and effects analysis (FMEA); potential problem analysis; reliability predictions; risk analysis; Kepner–Tregoe analysis; goal, question, metric (GQM); and quality functional development (QFD). These tools analyze design characteristics and identify product risks. Each of these tools can be studied further at the Web site http://www.isixsigma.com/, searching on a particular tool (e.g., QFD).

3. *Statistical tests*. Some key tools here: DOE, two-sample t-tests, analysis of variance, and control charts. These tests are looking for telling data patterns to pull the signal from the noise. They detect when there is truly a statistically significant difference occurring in the data versus simply random noise: that is, when there is truly an effect due to a variable rather than just random variation. The true effects are labeled "special cause" effects versus random noise, or what is called *common-mode variation*. Statistical testing prescribes adequate sample sizes to be taken from representative populations in a random manner and under proper control to determine whether the factor(s) being tested is significant at a given confidence level. For further study the reader is referred to a classic DOE book by Montgomery [4], where he quotes Mikel Harry, developer of six sigma, as saying: "Six Sigma is a process of asking questions that lead to tangible, quantifiable answers that ultimately produce profitable results."

The six sigma tool set, especially the collaborative and analytical tools, provide a good backdrop to ask questions without being intimidating or offensive. I clarified the questioning further to include questioning requirements, data, assumptions, process, and analysis. All of this questioning is conducive to learning and discovery and can greatly help the team bond together and achieve a common understanding of the design.

THE RASCI TOOL

Six sigma is a fact-based decision-making process. I have successfully facilitated and mediated a volatile (to the people involved) process problem that spanned several locations in the United States and Asia. This involved product specifications and documentation. These documents originated in a California laboratory and were then sent to a Colorado laboratory to add additional details. Then their final product was sent to the Asian manufacturing location to build product. Too many problems were arising, and each location did not want to be found as the "guilty party" behind those problems. In the six sigma mode, the operational data were simply collected and organized using a six sigma tool called RASCI (which is DFSS tool 2) to properly depict the operation and the responsibilities distributed across the several physical locations. RASCI is an acronym for:

Responsible: owns the problem or project

Accountable: represents the owner's accountability in signing off on work before it is effective

Supportive: can provide resources or can play a supporting role in implementation

Consulted: has information and/or capability necessary to complete the work

Informed: must be notified of results but need not be consulted

RASCI charted the level of responsibility at each step in the development process step. Emotion and defensiveness melted. The working group was able to focus and resolve the problem properly without undue emotion. The problems and process improvement opportunities became apparent to all participants as the RASCI tool was used. Defensiveness disappeared, much was accomplished, and good feelings prevailed.

DESIGN FOR SIX SIGMA

DFSS works systematically to optimize new designs, especially to meet a customer's needs and actually delight the customer. Optimizing the new design means:

- Minimizing defects toward the 3.4 defects per million opportunities or six sigma attainment
- Speeding up cycle time and reducing waste (lean)
- Minimizing product variation
- Managing variation (identifying variations and accommodating them in the design)
- Optimizing the design operating point for performance, stability, and robustness
- Ensuring flow down optimum tolerance allocations to subsystems
- Minimizing life-cycle costs
- Minimizing rework (six sigma: minimizing the "hidden factory" concept)
- Delighting customers and end users: all the downstream interfacing organizations
- Minimizing "dissatisfiers" such as high-impact failure consequences
- Incorporating product "delighters"
- Facilitating maintenance of the product delivered
- Smoothing product migration for customer adaptation of a new product
- Enabling future product enhancements ("perfective changes")
- Improving the underlying development process as gains are realized, which is one of the hallmarks of the six sigma process [i.e., improving the product (today's challenge), but also improving the underlying development process (tomorrow's products)]

DFSS has its own set of tools appropriate to each DFSS step. The major goal of the DFSS tools during development is to increase the comprehensiveness of the design team's thought. This is akin to pilots' navigational term *situational awareness*. A relevant example comes from a data storage company's experience. They had a picker mechanism in the data cartridge library which cost $2500 to manufacture. They were able to value-engineer this assembly and bring the manufacturing cost down to $1500. But the more interesting fact was that the people who cost-reduced the mechanism stated that if they had been involved at the beginning, they could have gotten the cost down to $600. Simpler designs usually are better all around.

Particular DFSS tools are used as appropriate and applicable to a given situation. There are several variations of the DFSS process, but the DMADV is an excellent approach. The reader can access the Web site http://www.isixsigma.
com/library/content/c020819a.asp for a detailed explanation and tool mapping for each DFSS phase.

1. *Define* the project goals and customer (internal and external) requirements.
2. *Measure* and determine customer needs and specifications, benchmark competitors, and industry.
3. *Analyze* the process options to meet customer needs.
4. *Design* (detailed) the process to meet customer needs.
5. *Verify* the design performance and ability to meet customer needs.

REQUIREMENTS DEVELOPMENT: THE PRINCIPAL CHALLENGE TO SYSTEM RELIABILITY

To quote Henry Ford: "If I had asked my customers what they wanted, it would have been faster horses." Developing and understanding product requirements for new products is a significant challenge. First, the customer typically has not thought through all the product features and needs. Functional requirements jump out first; other requirements seem to take time to discover. Then there are requirements of what the product should not do and how the product should behave when presented with off-nominal inputs (product robustness). Additionally, there are requirements to meet business and regulatory needs. Then there are questions of infrastructure, which often are not specified. Typically, they are simply assumed to be present and are noticed only by their absence.

Davy and Cope [1, p. 58] noted that "In 2006, C. J. Davis, Fuller, Tremblay, & Berndt [2] found accurately capturing system requirements is the major factor in the failure of 90% of large software projects, echoing earlier work by Lindquist [3], who concluded 'poor requirements management can be attributed to 71 percent of software projects that fail; greater than bad technology, missed deadlines, and change management issues.' " This requirements challenge has long been recognized and cited.

Note that the first three DFSS steps are focused on developing customer requirements and mapping the design solution to meet those needs. This is the major system development and reliability challenge. In this chapter we discuss two more of my favorite tools that have had the most impact, in addition to DOE and RASCI, discussed earlier.

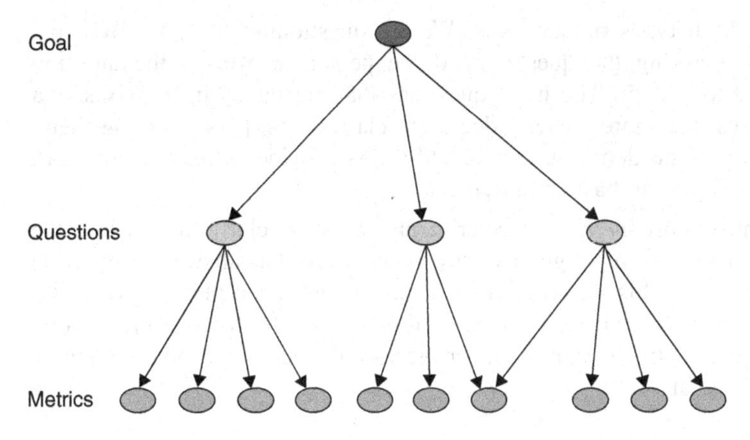

Figure 3 Tool 3: Goal–question–metric.

THE GQM TOOL

GQM is an acronym for "goal, question, metric" (Figure 3), an approach to metrics that was developed by Victor Basili of the University of Maryland–College Park and the Software Engineering Laboratory at the NASA Goddard Space Flight Center (http://en.wikipedia.org/wiki/GQM).

- Develop a set of project goals.
- Generate questions around those goals.
- Specify the measures that must be collected to answer those questions and track process and product conformance to the goals (i.e., in a measurable way).

These measures were to improve the development process by focusing on productivity and quality. This is a good introspective process tool. This author applies a variation of this tool to help define product requirements. Suppose that we begin with a critical set of requirements, say the one that has the most interest and impact: those requirements that are new, unique, or complex. Then we treat requirement as a goal. Next we brainstorm the questions we have around that goal. We use these questions to:

1. Clarify the meaning of the *goal* and find out what is really desired if that differs from what was stated originally. This may lead to a restatement of the goal (requirement). Mark Twain once said: "The difference between the right word and the almost right word is the difference between lightning and lightning bug."
2. Identify those *questions* whose answers will strengthen the stated requirement and validate its intent. GQM provides a good basis to

ask these types of questions. We are questioning to learn. Both the person asking the question and the person answering the question stand to benefit. The more questions that are raised in my six sigma classes, the more clever I deem the class or the person asking them. There are no dumb questions, and GQM provides a ready framework to question the basic requirements.

3. Identify those *metrics* or experiments or tests to clarify and answer the questions raised. A good metric is anything that answers a question of interest. This step typically amounts to an action plan for particular person(s) to study or analyze or test to provide the answer. A good outcome is to add conditions or increase the accuracy and definiteness of the goal.

THE MIND MAPPING TOOL

Mind mapping is a very visual process used to capture group thinking. It starts with a central theme. The one illustrated in Figure 4 is a real example of group thinking on improving the role of the IEEE Reliability Society's future initiatives. Ideas are recorded on the mind map as branches and leaves, as they are given. There is no judgment in the process; ideas are just captured. The facilitator can ask clarifying questions if needed to better understand the contributor's intention, but all suggestions are recorded. A mind map can capture notes, Web links, schedule management tasks, and more.

Mind maps were developed in the late 1960s by Tony Buzan as a way of helping students make notes that used only key words and images (http://en.wikipedia.org/wiki/Tony_Buzan). They are much quicker to make, and because of their visual quality, much easier to remember and review. The nonlinear nature of mind maps makes it easy to link and cross-reference different elements of the map. It associates ideas, which reinforces the learning and their memory. This process is a reflection of how the brain actually works. These graphic maps are more than two-dimensional since they can embed icons, files, sounds, and color all in one illustration.

I once asked the IEEE historian what characterized the great inventors of all time. The answer was that people like. Edison and DeVinci were very visual. A mind map works the same way that the great inventors work with its emphasis on images and association. Notes can be appended to mind maps to capture any rationale or insights revealed during the brainstorming process. The notes do not take up any visual space. They are represented solely by an icon that expands to reveal the associated notes when the cursor goes over the icon. This is like the embedded comment feature in Excel.

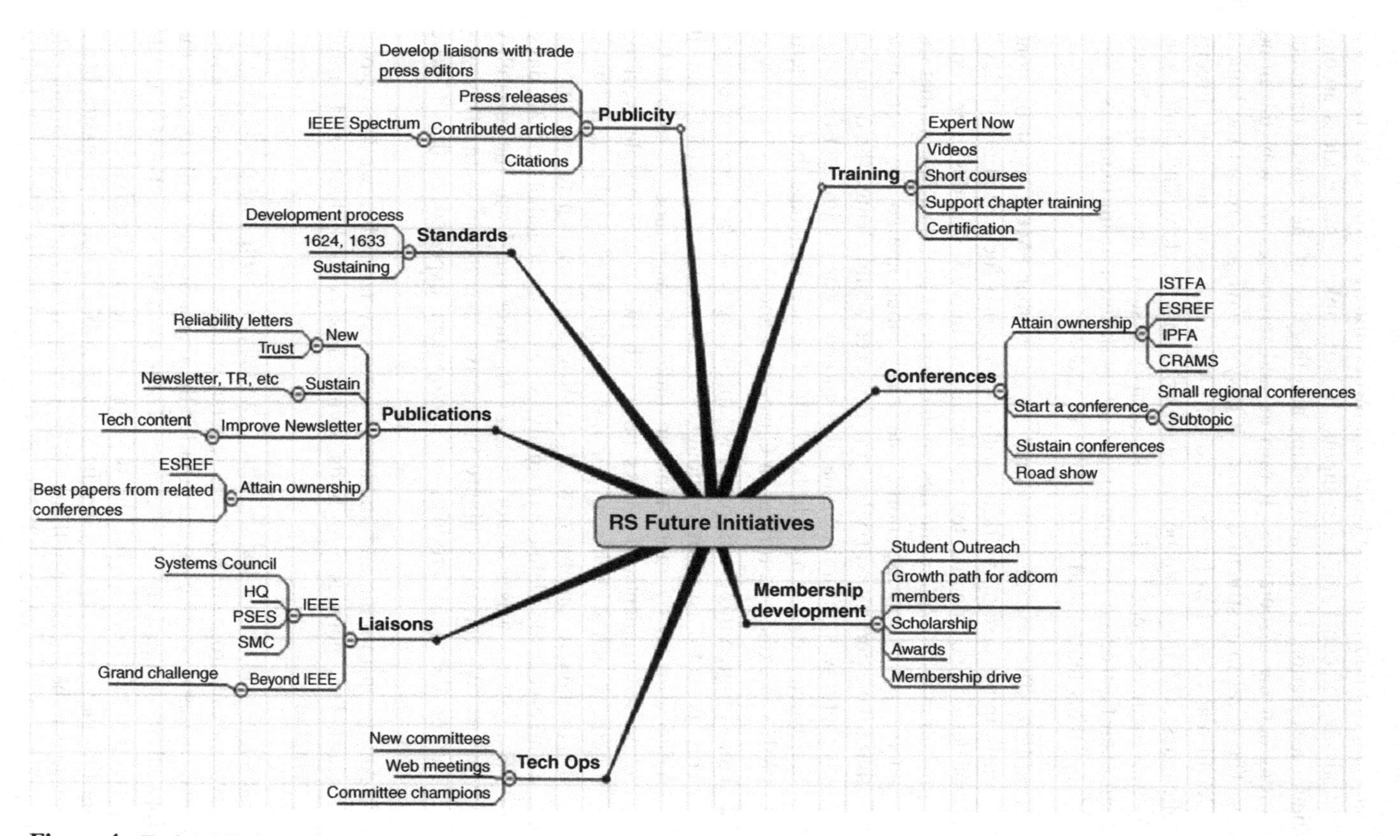

Figure 4 Tool 4: Mind mapping.

Building a mind map also builds teamwork among team members. It is a shorthand way of capturing team inputs. It is flexible to move ideas around for better association. Colors and linking and flagging of the related branches can strengthen the associations. It mirrors and reinforces the way the mind thinks. The mind map structure makes it easy to navigate around the logical map. Mind maps can then be exported into Microsoft tools and sent directly to colleagues. The maps can be truncated or expanded along any dimension, promoting focus as needed. The maps can also be stored in a common library and can even be worked on simultaneously by different team members.

More DFSS tools listed in my priority order:

1. FMEA. The critical operational aspect is that FMEA must be done early in the design, working with the designer focused at a high level of design and documented by the quality analyst. The designer's contribution is to explain the design and respond to the variations that the analyst posits: opens, shorts, late, early input, and so on. Every time that I have performed FMEA, the designer would change the design or say that it was unfortunate that this anomaly was not brought forth earlier, when the design could be changed to accommodate that problem. FMEA is explained further in Chapters 4 and 5.
2. Pair programming (http://en.wikipedia.org/wiki/Pair_programming). This is not customarily thought of as a six sigma tool, but it reduces errors by more than an order of magnitude. It also helps refine requirements. It works best with two senior people working together, each with his or her own complementary skill set. I used this technique on the new Hong Kong airport, where we finished a three-year-old delinquent analysis, working together for 30 days. My associate was a senior systems engineer and I was the reliability analyst. We formed a highly interactive, synergistic team. This was one of the most productive development experiences that I have ever had.
3. Any brainstorming and graphical capture system when working cross-team. This includes flowcharts, swim lanes, control charts, and timing diagrams.

Six sigma and design for six sigma can be energizing and productive for the development program. First, the reputation of six sigma successes across industry creates a positive expectation for the team. So they will buy into the various DFSS design analysis exercises. These exercises (e.g., RASCI) create an interactive environment for the development team. Creativity is fostered among the team and the analyst. The six sigma black belt's or master black belt's role is to be a teacher of the six sigma tools, which are used as applicable and appropriate to a project. The team shares and learns. The black belt is the process facilitator and has someone do the recording. The recording can be real time, with the six sigma tool being projected on a screen. There is always a "Wow" or discovery resulting from this systematic use of proven six sigma tools. It is a very effective and efficient analysis, discovery, and design documentation process. These six sigma analysis tools truly promote design completeness. The designers see this and feel their time was utilized profitably. It is a win–win situation for all.

REFERENCES

[1] Davis, C. J., Fuller, R. M., Tremblay, M. C., and Berndt, D. J., Communication challenges in requirements elicitation and the use of the repertory grid technique, *J. Comput. Inform. Syst.*, vol. 78, 2006.

[2] Davy, B., and Cope, C., Requirements elicitation—What's missing? *Issues Inform. Sci. Inform. Technol.*, vol. 5, 2008, p. 543.

[3] Lindquist, C., Required: Fixing the requirements mess: The requirements process, literally, deciding what should be included in software, is destroying projects in ways that aren't evident until it's too late. Some CIOs are stepping in to rewrite the rules, *CIO*, vol. 19, no. 4, 2005, p. 1.

[4] Montgomery, D. C., Design and Analysis of Experiments, John Wiley and Sons, 7th ed., 2009.

Chapter 9

Human Factors in Reliable Design

Jack Dixon

HUMAN FACTORS ENGINEERING

Human factors engineering (HFE) is a specialty engineering discipline that focuses on designing products and systems with the user in mind. The goal of HFE is to maximize the ability of a person or a crew to properly operate and maintain a product or a system by eliminating design-induced impediments and errors and to improve system reliability. It is important that the design engineer recognize the limitations as well as the capabilities and preferences of the human being in the system. A product that has been designed from this perspective will offer better customer satisfaction, require less training to operate and maintain, and will produce fewer errors and failures.

Human factors engineering is defined by MIL-STD-1908 [1] as "the application of knowledge about human capabilities and limitations to system or equipment design and development to achieve efficient, effective, and safe system performance at minimum cost and manpower, skill, and training demands. Human engineering assures that the system or equipment design, required human tasks, and work environment are compatible with the sensory, perceptual, mental, and physical attributes of the personnel who will operate, maintain, control and support it."

Human factors engineering as we know it today has its origin in World War II due to military needs to design and operate aircraft safely. Prior to that time, human beings were typically screened to fit the job or equipment rather than

Design for Reliability, First Edition. Edited by Dev Raheja, Louis J. Gullo.

having the equipment designed with humans in mind. Although human factors engineering grew initially in the defense and aerospace industries, it has since spread to all industries, including nuclear, space, health care, transportation, and even furniture design.

A DESIGN ENGINEER'S INTEREST IN HUMAN FACTORS

Since all products or systems are a collection of components that interact with each other and the external environment to achieve a common goal, the designer must be concerned about all the components and their interactions. Some of these components include humans. A human may be part of a system, a user of the system, or a controller of the system. Failures in products or systems are often blamed on human error. Quite often, human error is a symptom of something wrong with the design of the product or system. This may indicate a lack of consideration of the human aspect during the design phase of a product. Ideally, we want designers to design out all possibility of human errors. This is achievable with human-centered design practices.

Additional concerns with modern systems are the fact that the complexity of systems is increasing constantly, the technology is growing exponentially, and products and systems are becoming more difficult for people to use and understand. This makes it even more critical that humans be considered in product design. The reliability, safety, and usability of products and systems will be enhanced substantially if they are designed with people in mind. Safe and efficient operation of products and systems depends on properly designed and engineered interactions between the human and the machine.

HUMAN-CENTERED DESIGN

Human-centered design encompasses a wide variety of concerns. As the name implies, human-centered design places people at the center of design considerations rather than making people conform to the design. The design must accommodate humans' physical characteristics and mental processes such as perception and cognition. It must also take into account the environment in which a person must operate and the characteristics of the equipment, product, or system. The user also operates within an organizational framework that must also be considered. Other factors that influence human behavior, and should therefore be considered in the design process, include the technology being used, the management systems that are in place; and the procedures and processes under which the user will operate.

Role of Human Factors in Design

The product design should minimize human error and maximize human performance. Design considerations must include human capabilities, human limitations, human performance, usability, human error, stress, and the operational environment. Some typical topics that should be taken into account during the design process are shown in Table 1. This table is not meant to be all-inclusive; it is provided to highlight some major topics and to stimulate thought about what should be considered in a design.

With products and systems becoming more complex, organizations must be committed to human-centered design. They must fully integrate systems engineering with all types of specialty engineering, including human factors engineering; must develop system requirements to include user requirements; must test the product being developed with real users, assess the usability, and fix any shortcomings identified; and must use the tools and techniques that will facilitate this integrated design approach.

Hardware

Designers must consider the human being during the design process as related to hardware. Typical hardware-related considerations include weight, layout, access, anthropometrics, and ergonomics. Many examples of bad hardware design can easily be found in everyday products:

- Remember the Ford Pinto which required that the engine be dropped in order to change the spark plugs?
- Have you ever rented a car, pulled into a gas station to refuel it, and didn't know where the gas cap was? It could be on the right side, left side, or under the license plate. Inevitably, it is always on the side opposite the one nearest the pump when you pull in. Why aren't they always on the same side? Why aren't they always on the driver's side, as that's probably who will fill up the car?.
- Have you ever looked at a stove and wondered which knob controls which burner?

All these design flaws could have been eliminated if the designers had taken into account that essentially all products and systems have users who need to be considered and the design needs to accommodate them.

Software

Similarly, software design should consider the user. How many times have you installed a new computer program, opened it up, and discovered that you have

Table 1 Considerations in Human-Centered Design

Topic	Considerations
Displays	Can they be seen and understood by the user audience expected? Are they grouped in an intuitive way that makes sense to a user? Are they properly integrated with the controls Are the quality, relevance, and quantity of information presented appropriate? Are audio alerts appropriate for their intended function? Are warnings provided for an anomalous event?
Controls	Can they be operated by the user audience expected? Are they grouped for ease of operation? Are they labeled for easy understanding? Are they integrated properly with the displays? Is accidental activation prevented? Are emergency controls clearly identified?
Workspace design	Will a person fit comfortably? Can a person perform the assigned operations using the workspace provided? Is sufficient space provided? Are temperature, humidity, and ventilation controlled adequately? Is the lighting level correct for the functions to be performed? Are the viewing angles correct? Is the workstation eye height at the proper level? Does the workspace have adjustability?
Workplace environment	Will a person be comfortable in the environment? Is the lighting adequate? Are the temperature and humidity controlled properly? Is the background sound level suitable for the function being performed? Is the ventilation sufficient?
Maintenance	Has the product or system been designed with the maintainer in mind? Is the product modularized? Has proper access been provided? Are test points provided? Is diagnosis easy to perform?

no idea how to do anything with it—the icons are indecipherable, the order in which things need to be done is a puzzle, the color scheme makes the fonts unreadable, and no useful instructions are provided. What about Microsoft Windows' nonintuitive design ... Why would anyone design a system where one must go to the start button to shut off the machine?

Traditionally, human factors considerations in the design process have been focused on hardware aspects of the product or system. In the last couple of decades, more and more products have become more software intensive. The proportion of software to hardware in products has been increasing steadily. This trend has made development and change more rapid, and the greater use of software imparts a greater risk in both operations and maintenance. Software requires different skills; it makes diagnosis more challenging and generally imposes new considerations for the designer to think about during the design process.

Usability is a term used to describe the ease with which a person can employ a product or system. ISO 9241 [2] defines usability as the "extent to which a product can be used by specified users to achieve specified goals with effectiveness, efficiency and satisfaction in a specified context of use." Although the term can be applied to both hardware and software aspects, it is most often used in relation to software. As complex computer systems find their way into our everyday life, usability has become more popular and more widely utilized in recent years. Designers have seen the benefits of developing products with a user orientation. By understanding the interaction between a user and a product, the designer can produce a better, more widely accepted product. Desirable functionality or design flaws may be identified that may not have been obvious if human factors have not been considered. Implementing this human-centered design paradigm, the intended users of the product are kept in mind at all times. Maybe, then, the user won't have to go to the start button to stop the machine!

As with all good systems engineering, the most important part of the process is the up-front definition of requirements. It is critically important that complete user-interface requirements be identified early in the development process. There are numerous guidelines and style guides for user interfaces. An example of a guideline is ISO 9241, one of a series of guidelines for various aspects of computer–user interfaces. These provide a starting place for user-interface requirements generation, but they must be customized and tailored to fit the application.

Also important to implementing user interfaces successfully is design evaluation. This can be done using mock-ups or prototyping the user interface and testing with actual users. Requirements and implementations can then be adjusted in the design process when it is still cost-effective to make changes. Continuing user evaluation as the design evolves will ensure the best usability of the end product.

Human factors specialists do more than design friendly icons. They bring two important types of knowledge to bear on systems development: (1) human abilities and limitations, and (2) empirical methods for collecting and interpreting data from people. They define criteria for ease of use, ease of learning, and user acceptance in measurable terms. New technology demands much thought about the role of the tool. The distress and aversion that many people manifest toward computerization is perfectly rational in the presence of ill-conceived design. Too many of our systems confuse the operator.

One example of a human capability is time perception, which can be shown to affect the functional requirements of a software design. Table 2 shows time perception across various tasks and media. Transaction interactions should be without a perceived wait, and the standard deviation of all transactions should be less than 50% of the mean.

Human–Machine Interface

The interface between a person and a machine is of prime importance. The designer must be concerned with any and all parts of the product where a person must interact with the equipment. The human–machine interface becomes a critical component to be considered. This interface is defined as the plane of interaction between the person and the machine. It is across this plane that information and energy flow. The information is transferred across this interface from the machine to the person via displays and from the person to the machine via the controls. Therefore, displays and controls are of major importance in product or system design. Effective displays help to determine the proper action needed. Ineffective displays and controls contribute to errors that may lead to accidents.

Table 2 Perceptions Across Media

Human perception	Transaction time	Application	Preferred physical architecture
"Instantaneous"	Less than 1/3 second	Software development	Personal computer or workstation
"Fast"	Between 1/3 and 1 second	Simple query	Client/server
Pause	Between 1 and 5 second	Complex query and application launch	Thin client
"Wait"	Greater than 5 seconds	Action request	Background batch

The designer must also be concerned with the proper allocation of functions between person and machine, user and equipment. Machines are consistent; people are flexible. Machines are more capable of performing repetitive and physically demanding functions; people are more capable of performing functions that require reasoning. These are of prime significance when allocating functions to the machine or the person. This allocation of functions must ensure that the tasks assigned to each take into account what they do best, what their capabilities are, and what limitations they each have. Trade-offs between human and machine must be made regarding speed, memory, complex activities, reasoning, overload, and so on. In the next three sections we describe some of these considerations in greater detail.

Staff Requirements

The staff required to operate and maintain the system must be a design consideration. How many and what type of people will be needed to operate and maintain the system or product? Are properly qualified people available? Will training be required? If so, how much? Will these types of people be available in the future to support the entire life cycle of the system? Often, there are trade-offs that can be made during the design process that can reduce the number of people needed or the amount of training that will be necessary. For example, if the graphical user interface is intuitive to use, the time for the operator to learn to use the system can be greatly reduced.

Workload

The workload the system or product imposes on a user is a related concern. The tasks that a person must perform must be delineated. An assessment must be made of the amount of effort that each task will take both physically and intellectually. A matchup of the capabilities of the user to the tasks at hand must be ensured; otherwise, the user will become quickly dissatisfied with the product, or worse, will make errors as a result of task overload, which could lead to substantial undesirable consequences.

Personnel Selection and Training

Another important factor to consider during system design is who will be needed to operate and maintain the system. The proper people with sufficient skills and knowledge must be selected for the best fit to operate and/or maintain a system. Once selected, these people must be properly trained to do the job functions that have been allocated to them as a result of the system design process. Again, good product design can help to reduce these demands on the human operator or maintainer.

HUMAN FACTORS ANALYSIS PROCESS

Almost any technique used in system analysis can be applied to address the human element. However, there are numerous human factors–specific analysis techniques from which a designer can chose.

Purpose of Human Factors Analysis

The overarching purpose of human factors analyses is to develop a better, usable, and safe product or system. Various human factors analyses are conducted at different times in the development process and for different reasons. Analysis of human factors assists in the development of requirements, which is a critical step in the design process. As the product or system evolves, different analyses are conducted to define the human role in the system, to ensure that the human needs and limitations are being considered, to determine the usability of the product, and to confirm the ultimate safety of the system. Other analyses can be conducted to help guarantee customer acceptance of the product or system.

While the human factors engineer may be the lead for conducting the analyses, it should be a team effort. The team may vary depending on the stage of development or the particular analysis being conducted, but the results will always be better if it is a joint effort by a cross-functional team. Participants in the human factors analyses will always include the design engineer. The team may include other specialty engineers, such as system safety, reliability, software, and manufacturing engineers. Often, team participants may include management, marketing, sales, and service personnel. The team should be tailored to enhance the particular analysis being conducted.

Methods of Human Factors Analysis

As our products and systems have become more complex, it has become imperative that the old approach—either totally ignoring human considerations or making "educated guesses" based on intuition of how best to accommodate the human—be replaced by systematic analytical techniques to better match the human being and the machine. Although it is beyond the scope of this book to elaborate on all the various human factors analyses, Table 3 presents a sampling of the analyses available to a design team. The reader can find more detailed coverage of these and many other techniques in the books of Raheja and Allocco [3] and Booher [4].

Table 3 Human Factors Analysis Tools

Technique	Purpose	Description	Cost/difficulty	Pros	Cons
Prototyping	Addresses design and layout issues.	Mockup of user version.	Cheap if done early; more expensive as design progresses.	Quick way to show possible design before investing time and money on detailed development.	Can be expensive for complex systems or if the prototypes have to be updated each time a product or system changes.
Improved Performance Research Integration Tool (IMPRINT)[5]	Appropriate for use as both a system design and acquisition tool and a research tool. IMPRINT can be used to help set realistic system requirements; to identify user-driven constraints on system design; and to evaluate the capability of available staff and personnel to effectively operate and maintain a system under environmental stressors. IMPRINT incorporates task analysis, workload modeling, performance shaping and degradation functions and stressors, a personnel projection model, and embedded personnel characteristics data.	IMPRINT, developed by the Human Research and Engineering Directorate of the U.S. Army Research Laboratory, is a stochastic network modeling tool designed to help assess the interaction of soldier and system performance throughout the system life cycle, from concept and design through field testing and system upgrades. IMPRINT is an integrated Windows follow-on to the Hardware vs. Manpower III (HARDMAN III) suite of nine separate tools.	Moderately time consuming. Requires training. Inputs can be extensive.	Generates staff estimates and can be used to estimate life-cycle costs. Produces many types of reports.	Time consuming and a lot of data are needed.

(*continued*)

Table 3 (*Continued*)

Technique	Purpose	Description	Cost/difficulty	Pros	Cons
Task analysis	Used to analyse tasks and task flows that must be performed to complete a job.	A job/task is broken down into increasingly detailed actions required to perform the job/task. Other data, such as time, sequence, skills, etc., are included. This analysis is used in conjunction with other HFE analyses, such as functional allocation, workload analysis, training needs analysis, etc.	Moderate, but can get expensive for large, complex systems.	Relatively easy to learn and perform.	Task analysis was first developed for factory assembly-line jobs that were relatively simple, repetitive and physical, and easy to define and quantify. It is much more difficult to apply to complex or highly decision-based tasks.
Human reliability analysis	Used to obtain an accurate assessment of product or system reliability, including the contribution of human error.	Considers the factors that influence how humans perform various functions. It may include operators, maintainers, etc. The analysis is conducted using a framework of task analysis. First, the relevant tasks to be performed must be identified. Next, each task is broken down into subtasks and the interactions with the product or system are identified, and the possibility for errors is identified for each task, subtask, or operation. An assessment of the impact of the human actions identified is made. The next step is to quantify the analysis using historical data to assess the probability of success or failure of the various actions being taken. Also, any performance-shaping factors, such as training, stress,	Expensive and time consuming on large systems.	Thorough analysis of human errors and human–machine interactions.	Requires extensive training and experience in several disciplines.

		environment, etc., are taken into account. These factors may have an effect on the human error rate, which can be either positive or negative, but is usually negative. These might include heat, noise, stress, distractions, vibration, motivation, fatigue, boredom, etc.			
Job safety analysis	Used to assess the various ways in which a task may be performed so that the most efficient and safest way to do that task may be selected.	Each job or process is analyzed element by element to identify the hazards associated with each element. This is usually done by a team consisting of a worker, a supervisor, and a safety engineer. Expected hazards are identified and a matrix is created to analyze the controls that address each hazard.	Easy for simple jobs; more difficult for complex jobs.	Good for structured jobs.	Difficult if there is much variation in the job.
Technique for human error rate prediction	Used to provide a quantitative measure of human operator error in a process.	Developed during the 1960s for use in the nuclear industry for probabilistic risk assessment. It is a means of quantitatively estimating the probability of an accident being caused by a procedural error. The method involves defining the tasks, breaking them into steps, identifying the errors, estimating the probability of success/failure for each step, and calculating the probability of each task.	Can become expensive for processes with a lot of tasks.	Can be very thorough.	Obtaining good probability data.

(*continued*)

Table 3 (*Continued*)

Technique	Purpose	Description	Cost/difficulty	Pros	Cons
Link analysis	Used to evaluate the transmission of information between people and/or machines. It is focused on efficiency and is used to optimize workspace layouts and human–machine interfaces.	Links are identified between any elements of the system. The frequency of use of each link is determined. The importance of each link is then established. A link value is calculated based on frequency, time, and importance. System elements are then arranged so that the highest-value links have the shortest length.	Moderate	Operations and training can be enhanced. Safety critical areas can be identified.	Can become cumbersome on large systems.
Hazard analysis	Used to identify hazards and their mitigations during various phases of product/system development.	There are numerous types of hazard analyses, each with its own focus, and each conducted at specific times during the design effort. Preliminary hazard analysis is done early in the concept phase and helps provide information for conducting trade studies and/or developing requirements. Later, as the product development progresses, more detailed analyses are conducted: subsystem hazard analysis, system hazard analysis, and operating and support hazard analysis. Most often, all of these analysis techniques use a matrix format.	Moderate, depending on the size of the system being analyzed.	Provides a systematic way to determine hazards, assess risks, and present recommendations to mitigate or control.	Effectiveness and usefulness are dependent on knowledge (of the system and of hazards) of the team performing the analysis.

		This matrix (or sometimes individual hazard tracking sheets) typically contain the following information: hazard description, effect, operational phase, recommended controls, and risk assessments (both before and after controls are implemented).			
Fault tree analysis	Used to evaluate the likelihood of an undesirable event happening and identify the possible combination of events that could lead to it.	A deductive top-down analysis that identifies an undesirable event (top level) and determines the contributing elements (faults/conditions/human errors) that would precipitate it. The contributors are interconnected with the undesirable event, using a tree structure through Boolean logic gates. This construction continues to lower levels, until basic events (those that cannot be decomposed further) are reached. Combinations of these bottom-level basic events that can cause the top-level undesirable event to occur can be calculated (called cut sets). If failure rate data are provided, the probability of occurrence of the undesirable event can be determined.	Can get very expensive and time consuming for large systems.	Provides a graphical representation that aids in the understanding of complex operations and the interrelationships between subsystems and components. Can be used either qualitatively or quantitatively.	Significant training and experience are required.

HUMAN FACTORS AND RISK

Risk is inherent in all systems, and people add an additional dimension to the risk concern. Designers must be aware of both the risk posed by humans and the risk imposed on humans.

Risk-Based Approach to Human Systems Integration

The goal of all good systems engineering design is to reduce risk throughout the development cycle. This is accomplished by applying all relevant disciplines. In the past, however, the risks associated with human systems integration have often been ignored. Engineering risks are noticed at various times during the development due to implementation problems or cost overruns. Risks associated with human systems integration, however, are usually noticed only after a product or system is delivered to the customer. These end-state problems may lead to customer dissatisfaction and rejection of the product, due to it being too difficult or inefficient to use or, worse, they may lead to human error in the use of the product, which could have catastrophic consequences.

These operating risks can be traced back to failure to properly integrate the human needs, capabilities, and limitations at an early stage of the design process. Like all risk-reduction efforts, risk reduction in the area of human systems integration must be started early and continue throughout development of a product or system. This will ensure that requirements based on human factors are incorporated. It will allow design trade-offs to consider use of the product. This will produce a high level of confidence that the product or system will be accepted and usable by the user.

As has been emphasized throughout the book, the key to design success rests in the development of requirements. Requirements involving human factors are no different. A thorough and complete effort to specify human factors requirements in the earliest stages of product or system development will reap large rewards later. During development, other approaches to mitigating risk in the area of human systems integration include using task analysis to refine the requirements, and conducting trade studies, prototyping, simulations, and user evaluations.

HUMAN ERROR

It is easy to blame accidents and failures on human error, but human error should be viewed as a symptom that something is wrong within the system. Generally speaking, humans try to do a good job. When accidents or failures occur, they are doing what makes sense to them for the circumstances at hand. So, to understand human error, one must understand what makes sense to people. What "reasonable" things are they doing, or going to do, given the complexities, dilemmas, trade-offs, and uncertainties surrounding them? The designer must make the product or system make sense to users or they will be dissatisfied with it, failures will occur, or they will have accidents using it.

Designers often think that adding more technology can solve all human error problems. Quite often, however, adding more technology does not remove the potential for human error. It changes it and may cause new problems. Make sure that you know whether you are really solving the problem of human error by adding technology or are merely causing different problems. Take the simple addition of a warning light. What is the light for? How is the user supposed to respond to it? How does the user make it go away? If it lit up before and nothing bad happened, why should the user respond to it now? What if the light fails just as it is needed?

Many error-producing conditions that may cause a failure or an accident can be added to a product or system inadvertently. The designer must remain cognizant of the many things that can lead to human error, including environmental factors such as heat, noise, and lighting; confusing controls, inadequate labels, and poor training; difficult-to-understand manuals or procedures; fatigue; boredom; and stress. The goal is to eliminate those things that can contribute to human error.

Types of Human Error

Human error affects system reliability. There are many ways in which people can make errors. They can commit errors in calculations; they can choose the wrong data; they can produce products with poor-quality workmanship; they can use the wrong material; they can make poor judgments; and they can miscommunicate (just to name a few). Borrowing from Dhillon [6], we categorize human error using the design life-cycle perspective:

Design Errors These types of human errors are caused by inadequate design and design processes. They can be caused by misallocation of functions between a person and a machine, by not recognizing human needs and limitations, or by poorly designed human–machine interfaces.

Operator Errors These errors are due to mistakes made by the operator or user of the equipment design, and to the conditions that lead up to the error being made. Operator errors may be caused by improper procedures, overly complex tasks, unqualified personnel, inadequately trained users, lack of attention to detail, or nonideal working conditions or environment.

Assembly Errors These errors are made by humans during the assembly process. These types of errors may be caused by poor work layout design, distracting environment (improper lighting, noise level, inadequate ventilation, and other stress-inducing factors), poor documentation or procedures, or poor communication.

Inspection Errors These errors are caused by inspections being less than 100% accurate. Causes may include poor inspection procedures, poor training of inspectors, or a design being difficult to inspect.

Maintenance Errors These are errors made by maintenance personnel or the owner after a product is placed in use. These errors may be caused by improper calibration, failure to lubricate, improper adjustment, inadequate maintenance procedures, or designs that make maintenance difficult or impossible.

Installation Errors These errors can occur because of poor instructions or documentation, failure to follow the manufacturer's instructions, or inadequate training of the installer.

Handling Errors These errors occur during storage, handling, or transportation. They can be the result of inadequate material-handling equipment, improper storage conditions, improper mode of transportation, or inadequate specification by the manufacturer of the proper handling, storage, and transportation requirements.

When most people think of human error, the natural tendency is to think of an error by the operator or user. However, as can be seen from the categorization above, a major cause of human error that must be considered by the design engineer is design error. One must realize that during the design process, design flaws may be introduced. During the manufacturing process, assembly errors may occur. During quality inspections, product shortcomings may not be found due to human error. During subsequent maintenance and handling by the user, errors can occur. So the designer must be aware of all these types of defects and ensure that they are considered, and hopefully eliminated, during the design process.

Mitigation of Human Error

Mitigating risks due to human factors should begin early in the design process. Although it is always most desirable to completely eliminate conditions that lead to human errors, it is inevitable that human errors will occur. Therefore, error containment should be considered in the design. As technology has evolved, we have become more and more interconnected and the consequences of a failure, whether human-caused or not, can grow to enormous proportions due to this added complexity and interdependencies.

As an example, consider the August 14, 2003 failure of the power grid in the northeastern and midwestern United States and Ontario, Canada. The failure began when a 345-kV power line made contact with a tree in Ohio. Once the failure began, a chain reaction of events occurred due to numerous human errors and system failures. The failure propagated into the most widespread power failure in history, affecting 10 million people in Canada and 45 million people in eight U.S. states. The power outage shut down power generation, water supplies, transportation (including rail, air, and trucking), oil refineries, industry, and communications. The failure was also blamed for at least 11 deaths.

As products and systems continue to become more complex and interwoven into our culture, designers must not only ensure that human–machine interfaces are understood and usable, but must also consider the potentially far-reaching consequences that a failure might have, and design products and systems in a way that will minimize these consequences.

DESIGN FOR ERROR TOLERANCE

Error-tolerant systems and interfaces are design features worthy of consideration. Error-tolerant systems minimize the effects of human error. Error tolerance capabilities added to a system improve system reliability. Human error is frequently blamed for accidents, especially high-consequence accidents. It is not uncommon to hear 60 to 90% of accidents attributed to human error. While we should be eliminating opportunities for human error to occur, we should also be designing products to be error tolerant. Designers need to consider the consequences of failures of their designs. Errors lead to consequences; consequences should be minimized or eliminated. This is the essence of error-tolerant design. A design that tolerates errors avoids the consequences. By providing feedback to the user on both current and future consequences, compensating for errors, and providing a system of intxelligent error monitoring, a design can be made to be more error tolerant. One should note that the emphasis is on "intelligent" monitoring and feedback. The system should not just provide an indecipherable error message, such as "ERROR #404." This is the type of error message so often seen these days by the average computer user, who has no idea what the problem is or how to fix it. The system should provide a more useful message that describes the problem and provides useful information to the user about the error and what should be done to resolve it.

CHECKLISTS

Checklists can be an effective tool to help a designer and/or a user. There are two main types of checklists. One is to provide the designer with a checklist to follow during the design and testing of a product or system. The second is to provide the user with a checklist to follow when using the product or system so that nothing will be omitted.

A human factors design checklist is typically a long list of design parameters that should be considered during the product design process. These lists can be based on previous experience, lessons learned, and/or some published design guide. For example, a checklist derived from MIL-STD-1472 [7] can be used to help a design engineer ensure that a design will be usable by people for its intended purpose. It should be realized that checklists have limitations. It is impossible for a checklist to cover all variables and combinations of conditions for all designs. Despite that shortcoming, checklists provide guidance to the designer for things that need to be considered. They can also serve as tools for test engineers to verify that the product or system has been designed and produced with the user in mind.

Checklists are created for a user to ensure that the product or system is operated correctly, that all procedures are followed, and that uncertainty in operation is avoided. A well-known example of a checklist is the preflight checklist, used by pilots prior to take off.

TESTING TO VALIDATE HUMAN FACTORS IN DESIGN

Human factors validation is just as important as the development of adequate requirements for product or system specification. It is important to test the product or system against each human factors requirement and to verify that the requirement has been met adequately. Human performance requirements should be validated in system test plans and demonstrated in usability tests, and the results addressed in test reports. The product or system should be tested by representative users to verify that it functions as planned and can be operated properly and safely by the intended user.

REFERENCES

[1] *Definitions of Human Factors Terms*, MIL-STD-1908, U.S. Department of Defense, Washington, DC, Aug. 1999.

[2] *Ergonomics of Human System Interaction*, ISO 9241-11, International Organization for Standardization, Geneva, 1998.

[3] Raheja, D. G., and Allocco, M., *Assurance Technologies Principles and Practices: A Product, Process, and System Safety Perspective*, 2nd ed., Wiley, Hoboken, NJ, 2006.

[4] Booher, H. R., *Handbook of Human Systems Integration*, Wiley, Hoboken, NJ, 2003.
[5] U.S. Army, *MANPRINT in Acquisition: A Handbook*, U.S. Department of Defense, Washington, DC, Apr. 2000.
[6] Dhillon, B. S., *Design Reliability: Fundamentals and Applications*, CRC Press, Boca Raton, FL, 1999.
[7] *Human Engineering*, MIL-STD-1474F, U.S. Department of Defense, Washington, DC, Aug. 1999.

Chapter 10

Stress Analysis During Design to Eliminate Failures

Louis J. Gullo

PRINCIPLES OF STRESS ANALYSIS

Stress analysis begins by identification of those values or parameters that stress a design during any particular user application, mission profile, or operational profile. Once the values or parameters are identified, they must be controlled in the design. Components and materials should be selected that have the strength to withstand the stress parameters in the nominal case as well as the worst-case operating conditions.

Failures during production and early life wearout mechanisms are precipitated by stress. When stress exceeds the strength of a design, parametric degradation and failures occur. Design for reliability is achieved by ensuring that the strength of the design has adequate margin to handle the stresses applied to an item over its lifetime. With adequate design margin on those parameters, the life of a product or system is extended through elimination of design weaknesses attributed to early life wearout failure mechanisms and avoidance of failures that are critical to mission success. Electrical and mechanical stress analyses are performed to determine the operating stresses to be experienced by each component, product, and system commensurate with the reliability prediction calculations and the engineering design information available. Thermal stress analysis may be carried out on components that are sensitive to high and low temperature extremes, rapid thermal slew rates, continuous thermal cycling, and chemical reactions such as those involving oils, lubricants, and tires.

Design for Reliability, First Edition. Edited by Dev Raheja, Louis J. Gullo.

MECHANICAL STRESS ANALYSIS OR DURABILITY ANALYSIS

Mechanical stress analysis assesses the ability of equipment to withstand the stresses imposed by operational use, maintenance, shipping, storage, and other activities throughout the product life cycle. The mechanical stress analysis or durability analysis process should be capable of evaluating the simultaneous long-term effects of mechanical, thermal, vibration, humidity, and electrical stresses. Durability analysis or mechanical stress analysis may be a structured process that includes the following major steps:

1. Determine operational and environmental stresses that the equipment will experience throughout its lifetime, including normal and worst-case operational profiles, mission profiles, shipping or transportation profiles, handling, storage, and maintenance.
2. Determine the magnitudes and specific locations within a design where significant stresses and failures are likely to occur, using, for example, design failure modes effects analysis and finite element analysis.
3. Determine how long the significant stresses can be withstood or sustained using the appropriate accumulative fatigue or damage models (e.g., Arrhenius equations, inverse power laws).
4. Report the results as a list of failure modes, failure mechanisms, and failure causes that are prioritized and ranked according to their failure criticality or the time expected for failure to occur, such as failure probability or failure rate.

FINITE ELEMENT ANALYSIS

Finite element analysis (FEA) is a numerical technique used by engineers to solve specific problems of stress and other engineering problems. FEA is also used by mathematicians for finding approximate solutions to partial differential and integral equations. The solution approach is based either on eliminating the differential equation completely (steady-state problems), or approximating the partial differential equation with a system of ordinary differential equations, which are then integrated numerically using standard techniques. FEA is beneficial for solving partial differential equations when the domain changes during a solid-state reaction with a moving boundary, or when the desired precision varies over an entire domain. FEA is useful in performing stress analyses and solving problems involving wearout failure mechanisms. FEA solves thermal stress problems as well as mechanical loading problems. FEA can calculate and plot thermal, random vibration, and mechanical shock distributions. Based on the characteristics of these distributions, failure predictions and acceleration factors of wearout mechanisms can be ascertained. There are other methods that involve parallel solutions for FEA using Monte Carlo simulation. Solution strategies for parallel implementation have been developed in connection with the Monte Carlo and weighted integral methods to produce efficient numerical handling of FEA for two-dimensional plane stress–strain problems. Monte Carlo used in conjunction with the local average method is also used to extend FEA to three-dimensional solid structures. Monte Carlo has the advantage of providing accurate solutions for any type of problem whose deterministic solution is known either numerically or analytically; however, its disadvantage may be the high computational effort that is required to solve certain complex probabilistic problems. FEA with or without Monte Carlo analysis is used extensively in determining the physics-of-failure models of various electronic assemblies and components. Monte Carlo analysis is used to develop Monte Carlo models. Monte Carlo modeling is discussed in Chapter 4.

PROBABILISTIC VS. DETERMINISTIC METHODS AND FAILURES

The following are the differences between probabilistic analysis and deterministic methods:

- A deterministic method results in a definite conclusion that a failure does or does not exist after a single trial.
- A probabilistic method results in a probability that a failure does or does not exist after a number of trials.

The following are the differences between probabilistic and deterministic failures:

- Deterministic failures are normally found in a single test process and do not require repeated testing of the same functions over a period of time.
- Probabilistic failures require repetitive testing of the same functions to be detected over a period of time.

HOW STRESS ANALYSIS AIDS DESIGN FOR RELIABILITY

The basic premise in designing for reliability is that $P[\text{fail}] = P[\text{stress} > \text{strength}]$. The stress and strength variables (or vectors) are design parametric values derived from component device manufacturer ratings in datasheets and empirical data measured or simulated from the actual application. You could even simulate P[fail] for a single component from random samples of stresses and strengths (at failures). The *parts stress analysis method*, a reliability prediction method defined in MIL-HDBK-217 [1], is used during the system or product development process. The method, which assumes that component failure modes are probabilistic, calculates component failure rates used to calculate design reliability.

The parts stress analysis method, which assumes that the times to failure of all components are distributed exponentially, resulting in constant failure rates, is performed when the detailed design is stable and firm, in order to determine component failure rates used to calculate product and system failure rates. The determination when the design is stable and firm is based on (1) the availability of the engineering design information; (2) the maturity of several design artifacts, including the parts list or bill of materials, electrical circuit schematics, product specifications, and assembly drawings; and (3) an understanding of circuit performance. Circuit performance understanding improves with analytical and empirical knowledge of performance, which replaces the assumptions. Assumptions of performance are reduced through accuracy of models, simulations and measured parameters in the application of the components, their stresses over the operating profiles and environmental conditions, their design margin and derating, their quality factors, and their expected environments. These factors should be known when the design is stable, or at least there should be a way of determining these factors based on the state of the hardware definition. Failure rates for common components can be modified by the appropriate factors (e.g., pi factors for each component model in MIL-HDBK-217) to account for the effect of applied stress. Where unique components are used, any assumptions regarding their failure rate factors should be identified by experimentation, simulation, or testing methods. The parts stress analysis method is the most accurate method of reliability prediction prior to measurement of reliability under actual or simulated use conditions.

Measurements of reliability may be performed by experimentation, modeling and simulation, testing, or field experience (performance history in customer use applications). Measurements of this type are usually more accurate than the parts stress prediction method, and because of this, may be used to update and replace the previous predictions performed using the parts stress method.

DERATING AND STRESS ANALYSIS

Derating is the selection and application of components and materials for the parametric strength inherent in their designs so that the impact of stress exposure is minimized by ensuring that the actual applied stress values are less than the rated values for each specific application. Typical electrical stress values or parameters applied to an electrical design are power, current, frequency, and voltage. These stresses are important parameters in the design since they affect inherent device failure rates. Derating decreases vulnerability to the electrical, thermal, and mechanical stress values or parameters of a design or increases the electrical, thermal, and mechanical strength of a design. Whichever your perspective, the end effect is the decrease in the actual worst-case electrical, thermal, and mechanical stress percentage when comparing the applied parameter to the rated parameter. For example, an electrical stress percentage is the ratio of an applied maximum stress over the maximum manufacturer's rated design strength, such as applied peak voltage divided by rated peak voltage.

The greater the difference between strength values and stress values, the greater the *factor of safety*, also known as the *design margin*. With a substantial design margin, through elimination of design weaknesses, zero mission-critical failures are possible. An adequate design margin may be achieved by changing the design so that strength is added, through components or material selection, or stress is removed by changing application parameters.

STRESS VS. STRENGTH CURVES

Figure 1 shows a possible relationship between the product strength and the product stress. Each curve represents a distribution of a particular design value or an accumulation of design parametric values. If the overlap between the two distributions is virtually eliminated when the strength curve moves farther to the right, or the stress curve moves farther to the left, as seen graphically in Figure 1, the probability of failure approaches zero. As the strength curve moves right and the stress curve moves left, such that the positions of the

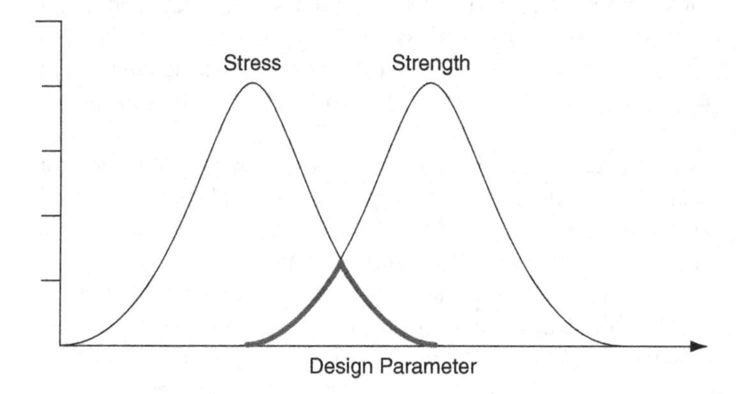

Figure 1 Stress–strength overlap.

strength and stress curves are reversed in Figure 1, and continue to separate further, moving in opposite directions, the probability of failure approaches 1. For most cases, a probability of failure that approaches zero over a certain period of time and under certain environmental conditions is evidence of an adequate design margin.

Contrary to popular opinion, the overlap of the two distributions (stress curve and strength curve) does not represent the probability of failure. Overlap indicates only that there is positive probability of failure. When there is total overlap of equally shaped curves, the probability of failure is 0.5 rather than 1. The probability of failure becomes 1 only when the stress curve moves completely to the right of the strength curve (i.e., when there is no overlap).

Example 1

Selecting Parameters That Can Be Derated for a CMOS Device

CMOS devices typically have negligible quiescent power consumption compared to the power dissipation during switching. The power dissipation (P) of a CMOS device is given by the equation

$$P = (C_{\mathrm{pd}} + C_{\mathrm{L}})V_{\mathrm{CC}}^2 F + V_{\mathrm{CC}} I_{\mathrm{CC}} \tag{1}$$

where

C_{pd} = power dissipation capacitance
C_L = load capacitance
F = switching frequency
I_{CC} = quiescent supply current
V_{CC} = supply voltage.

In this example, the CMOS device manufacturer's datasheet specified a rated power dissipation of 170 mW at 6 MHz, $C_{\mathrm{pd}} = 50$ pF/buffer, and I_{CC} (maximum specified, for $V_{\mathrm{CC}} = 6\mathrm{V}$) = 160 μA. The load capacitance (C_L) is assumed to be 50 pF, which is the value used for test conditions in the datasheet.

The power dissipated by the device is the power that is lost as heat and is not the same as the device's output power. For CMOS devices, all the power drawn by the device is dissipated, because the output current is negligible. For certain CMOS device datasheets, output power may be the only power parameter published. The rated power dissipation will have to be calculated from additional data obtained from the manufacturer's datasheets. This power calculation should not be confused with the calculation for actual power dissipated. The reader should also be aware that some manufacturers specify a slightly different equation for calculation of CMOS power dissipation stress, which is valid for a 50% duty cycle:

$$P = \frac{1}{2}CV^2F \tag{2}$$

Determine the actual parametric value for power dissipation.

The circuit application requires 15 buffers to be wired. The value of C_{pd} is 750 pF (15 × 50 pF). From the datasheet values and application conditions, and using equation (1) for the power dissipation:

$$P = 0.0288 \times 10^{-6}\ \mathrm{F} + 960 \times 10^{-6}\ \mathrm{W} \tag{3}$$

$$P = 173.76\ \mathrm{mW}(\text{at frequency} = 6\ \mathrm{MHz})$$

Considering that the rating specified for the power is 170 mW, comparing the actual power value to the rated power value, the actual power is overstressed (>100% stressed). If the frequency were derated by 10% (frequency = 5.4 MHz), the power would become 156.48 mW, which is below the rated power. The power dissipated could also be given a 10% derating so that the power would be lowered even further, down to 153 mW, by decreasing the operating frequency below 5.4 MHz. However, decreasing the operating frequency may not be feasible, given the circuit design constraints. The variability and limitations of the operating frequency are highly dependent on each circuit application, timing analysis, and the performance requirements.

Following are five other examples of stress analysis and derating methodologies.

1. A table is designed to handle 10 lb of continuous load over 5 years. A stress of 100 lb will be too much for the strength of the table and will cause damage 100% of the time. Loads of 20 lb will not cause immediate damage, but with the accumulation of fatigue, failure of the table will occur prior to the end of its 5-year life. An applied maximum stress load requirement of 9 lb may be applied to the table design, which provides a 10% derating and ensures continuous load-bearing capability over a 5-year period with a high degree of confidence.

2. A linear power supply is designed for a 100-A load at 12-V. If a 200-A load is applied, the power supply operation may become unstable and fail within 40 h when a continuous load is maintained. If a 250-A load is applied, the power supply may fail instantly, with a 90% degree of certainty. As an example, if the load is derated to 80 A, providing a 20% derating (20 A/100 A), the power supply may last for 10 to 20 years, notionally, under continuous loading. The determination of a power supply's lifetime depends on the average and peak currents loaded throughout its lifetime. A switching power supply may last 20 to 30 years given the same design derating and loading as that of the linear power supply. Comparing the linear supply to the switching supply, this increase in life is due to the duty cycle, in which the components are turned on and off very fast to generate the voltage, current, and power output. The switching design means less stress on the components in the design topology compared to the linear design topology. This reduced power stress is the reason for 50% reduction in the power for CMOS, as shown in equation (2), when comparing switching integrated circuits to linear integrated circuits.

3. An integrated circuit (IC) has a 10 : 1 fan-out rating. *Fan-out* indicates the number of device inputs that connect to a single device output as well as the number of interconnects for devices connected to the same signal, or load, or data output. Good fan-out means that the ratio of inputs to output is less than the rated fan-out and that the actual current drain on a device being analyzed is well below the maximum current rating for a device's output. If 20 device inputs are connected to a particular device's output with a rated fan-out of 10, this causes excessively high current drain above the particular device's specified rating, increases power dissipation, and results in overstress to the IC and device failure. With the increased power dissipation above the specified ratings, transistor junctions and gate stresses occur with performance degradation (soft failure) or hard failure (permanent damage) through tunneling current and physical fatigue. If eight devices are connected to the single output, providing a 20% derating, the output will provide continuous operation for a 20-year period, as an example, and a failure rate of 1 failure per million part hours. The lower the fan-out, the higher the probability of successful device operation over the device's specified lifetime (e.g., 20 years).

4. A laptop computer is specified to operate over a 10-year life in a ground-benign environment. The customer requirements do not specify a duty cycle, power cycle, or operating profile. The manual that describes operation of the laptop states that a power management function is provided and is user configurable. Stress to the laptop is caused by excessive power switching on/off, and by running at high power continuously. The user could subjectively derate the laptop by avoiding power cycling more than once per month, by running a power standby mode when the laptop is not in use, and by ensuring that the laptop remains in an ambient environment controlled for room temperature on a shock-isolated minimized-vibration surface at 50% humidity.

5. Digital logic devices called transistor–transistor logic (TTL) experienced a significant design reliability improvement in the 1980s when the Schottky TTL was developed. Schottky TTL advanced past normal TTL with improvements in device failure rates by incorporating clamping diodes in the topology of the logic devices. The diode clamping across the transistor's collector and base prevents transistor saturation, which means that the collector's junction is prevented from total forward biasing by limiting the voltage across the collector and base (V_{cb}) to 0.55 V or less. The stress of the junction's switching transients (from the ON state to the OFF state, and vice versa) is minimized, and the strain of the junction's recovery from ON to OFF is lowered since the junction is not totally saturated while in the ON state. Diodes on the inputs to logic devices also prevent damage to the junctions from electrostatic charges.

One particular reference MIL-HDBK-338 [2], paragraph 5.2.1.7, describes "derating" as follows:

Derating

Derating may be defined as the practice of limiting electrical, thermal, and mechanical stresses on parts to levels below their specified or proven capabilities in order to enhance reliability. By derating, the margin between the operating stress level and the maximum stress level for a part is increased, thus providing added protection from system anomalies unforeseen by the designer. Some suggested derating guidelines for microcircuits are outlined in the following table.

	Environmental derating factor	
Parameter	Benign	Severe
Supply voltage[1]	0.80	0.70
Input voltage	0.70	0.60
Output current	0.80	0.70
Maximum junction temp. (°C)	105	80

[1]For devices with dynamic supply voltage ranges only; all others use the manufacturers' recommended supply voltage. Derating below 80% of the supply voltage may operate the device below recommended operating voltages.

Another reference worth mentioning is IEC 61709 [3]. This international standard is intended for the reliability prediction of components as used in equipment and for organizations that have their own field failure data, describing how to use those data to perform reliability predictions. It can also be used to allow an organization to create its own failure rate database, and describes the reference conditions for which field failure rates should be stated. The reference conditions adopted in this standard are typical of the majority of applications of components in equipment. Using the stress models presented allows extrapolation of failure rates to other operating conditions, which in turn permits the prediction of failure rates at assembly level. This allows for the estimation of component reliability based on the effects of design changes or changes in the operating use profile, such as changes in the environmental conditions. The stress models in IEC 61709 are generic. This international standard does not contain failure rates but describes how they can be stated and used. This approach allows a user to select the most relevant and up-to-date failure rates for prediction from a source that they select. This standard also contains information on how to select the data that can be used in the models presented.

SOFTWARE STRESS ANALYSIS AND TESTING

Software stresses may be analyzed and tested by executing the software code with use cases or test cases that are based on expected operational scenarios or accelerated stress exposures. Software may be stressed by exposure in test cases with more than the maximum anticipated loads from use cases, with no loading at all or with the appropriate load from a use case in a very short time frame, such as an accelerated use case. Software is also stressed in an accelerated use case by running the software longer than the specified runtimes. Systems fail when arrays or files run out of allotted space or the physical capacities of tapes, disks, or buffers are exceeded. Short-term buffer capacities may be exceeded or temporary arrays may overflow. Memory leakage occurs when software allocates memory for a specific function but the software does not release the allocated memory back to the operating system after the function is completed. A memory leak degrades the performance of the system by reducing the amount of memory available. As more of the available memory is allocated, the system fails to work correctly, applications fail, or the system slows down. This type of failure is a major problem in real-time data acquisition and control systems or in batch systems with fixed time constraints. Software stress analysis is conducted to identify weaknesses in the software structure which negatively affect software structural integrity.

Software design metrics are useful in evaluating software stresses. Three metrics are used to extract design information from software code to identify structural stress points in a system being analyzed. The first software design metric is Di, which is an internal design metric which incorporates factors related to a module's internal structure. Di is a design primitive to assess design cohesion. The second software design metric is De, which is an external design metric. De focuses on a module's external relationships to other modules in the software system. De is a design primitive to assess design coupling. The third and final metric is D(G), which is a composite design metric. D(G) is the sum of Di and De.

Since stress point modules generally have a high probability for being fault-prone, engineers use the software design metric information to determine where additional testing effort should be spent and determine what modifications are needed. A study was conducted using software design metrics on part of a distributed software system, written in C, with a client–server architecture, and identified a small percentage of its functions as good candidates for fault-proneness. Files containing these functions were then validated by the real defect data collected from a recent major release to its next release for their susceptibility to faults. Normalized metrics values were also computed by dividing the Di, De, and D(G) values by the corresponding function size determined by nonblank and noncomment lines of code, to study the possible impact of function size on these metrics.

A framework should be used to provide software design analysis and testing guidance through a stress-point resolution system, which can be based on a software module signature for software module categorization. A stress-point resolution system includes stress-point identification and the selection of appropriate mitigation activities for the stress points identified. Progress has been made in identifying stress points to target the most fault-prone modules in a system by the module signature classification technique [4].

STRUCTURAL REINFORCEMENT TO IMPROVE STRUCTURAL INTEGRITY

The purpose of stress analysis and stress testing is to identify high stress points in the design, to design-out faults, and to reinforce the structure and strength of the design, which improves structural integrity and design reliability. The selection of appropriate software stress mitigation activities to remedy identified stress points is called *structural reinforcement*. It is through structural reinforcement of the software code using software design metrics that the integrity of the software structure is achieved. As software structural reinforcement is accomplished, in concert with hardware structural reinforcement through stress analysis and stress testing, a system reaches structural integrity and improves system reliability.

REFERENCES

[1] *Military Handbook for the Reliability Prediction of Electronic Equipment*, MIL-HDBK-217, Revision F, Notice 2, U.S. Department of Defense, Washington, DC, Feb. 1995.
[2] *Electronic Reliability Design Handbook: Design Assurance Guidelines*, MIL-HDBK-338, U.S. Department of Defense, Washington, DC, 1984.
[3] *Electronic Components—Reliability—Reference Conditions for Failure Rates and Stress Models for Conversion*, IEC 61709, International Electrochemical Commission, Zenevz,
[4] Zage, D., and Zage, W., A stress-point resolution system based on module signatures, in *IEEE Proceedings, 28th Annual NASA Goddard Software Engineering Workshop*, Dec. 2003.

Chapter 11

Highly Accelerated Life Testing

Louis J. Gullo

INTRODUCTION

Highly accelerated life testing (HALT) is a method used for rapid acceleration of the precipitation and detection of failure mechanisms, latent manufacturing defects, and design weaknesses over time. The failure acceleration occurs through the application of a combination of environmental and electrical stress conditions, such as temperature, vibration, humidity, power, and voltage. HALT and highly accelerated stress testing (HAST) apply accelerated stresses that are applied nonuniformly in varying stress combinations, called *step stresses*, and various environmental conditions, called *load case conditions*.

Gregg Hobbs, who pioneered the HALT process and coined the term HALT, states: "In HALT, every stimulus of potential value is used under accelerated test conditions during the design phase of product in order to find the weak links in the design and fabrication processes. Each weak link found provides an opportunity to improve the design or the processes, which will lead to reduced design time, increased reliability, and decreased cost" [1].

Traditional reliability tests such as accelerated reliability testing (ART) and accelerated life testing (ALT) use environmental stresses applied uniformly and consistently to expose failures and to develop acceleration factors to equate the stresses applied in a test to reliability or to life in the actual customer use environment. With HALT, acceleration factors are difficult if not impossible to determine, and test time cannot be correlated to customer use time. The

Design for Reliability, First Edition. Edited by Dev Raheja, Louis J. Gullo.

reason that acceleration factors are difficult to develop for HALT data is due to the rapid acceleration of stress conditions and the combination of stress conditions where existing acceleration models are not adequate. This means that test times accumulated in HALT under step stress conditions may not be correlated to customer use times, since the highly accelerated stress levels in HALT are stepped, not steady-state or constant stress levels. For example, the Coffin–Manson equation is used to calculate the acceleration factor from a constant temperature cycling test, which is used to correlate the stress test time to actual application time or failure rates in customer environments. The Coffin–Manson equation would be useful in HALT for a single temperature-cycle step stress condition, but not for the entire temperature-cycle load case, composed of multiple step stresses. HALT focuses on design enhancements and refinements, such as design changes to achieve a reliability improvement instead of focusing on acceleration factors and predicting or demonstrating field reliability. Therefore, HALT does not result in a metric for calculating reliability, such as mean time between failures (MTBF).

HALT is a series of individual and combined load cases and step stresses that are applied in steps of increasing intensity until the product fails or the preestablished limits are reached. Predefined stress limits are selected when the destruction levels are known from previous tests, or when fundamental limits of the technology are known through supplier specifications. Where testing to destruction limits is not necessary, testing may be called highly accelerated stress testing (HAST) instead of HALT. HALT or HAST stresses are typically some combination of the following: single-, double-, or multiaxis random vibration (electrodynamic or repetitive shock), high-temperature step stresses, low-temperature step stresses, temperature cycling (or thermal shock), power cycling, voltage variation, four-corner tests, and humidity. Four-corner tests are testing for two stress conditions simultaneously where stresses are applied at the upper and lower boundaries for each stress, resulting in four stress steps. For example, a four-corner test for stress A and stress B results in a test at the upper limit of stress A with the upper limit of stress B, upper limit of stress A with the lower limit of stress B, lower limit of stress A with the upper limit of stress B, and finally, the lower limit of stress A with the lower limit of stress B.

Humidity is valuable as a stress condition only if moisture exposure is a concern for the product design. Any stress or any stimulus that is deemed product sensitive should be employed. By comparing the stresses to the design strength, measurable design margins are determined. The design margin is also known as the *factor of safety* (FOS). FOS is the fraction of design capability over the design requirement. There is a direct correlation between the FOS and the measure of the reliability of a particular design. The correlation is a directly proportional relationship, such that as the difference between the stress and the strength increases, the predicted reliability over the life cycle increases. Although this correlation exists, there is still no way to estimate the mean life or mean time to failure (MTTF) or MTBF of the product as a result of the HALT process. To accomplish this, a reliability test such as a reliability demonstration test or reliability development growth test (RDGT) is needed.

RDGT and test analyze and fix (TAAF) tests are similar to HALT but were developed for military products and systems many years before the HALT process development. RDGT provides a quantifiable reliability metric, such as MTBF. Its results may be useful in driving design changes to meet specified reliability requirements. HALT does not replace these types of military reliability tests but, rather, complements these test approaches. Several differences appear in a comparison of the HALT and RDGT approaches. One difference is that HALT finds design weaknesses more quickly than does RDGT. This is accomplished through highly accelerated environmental test methods that provide time compression, which results in precipitating failure mechanisms over a shorter period of time than with RDGT or TAAF. When given a choice, system or product developers should adopt a two-stage reliability test approach, involving reliability demonstrations and HALT. Both approaches have value and offer benefits that complement each other.

Next we explore HALT in detail by examining what HALT is and what it is not, and discuss time compression and test coverage. To understand what HALT is, let us explore multiple factors in end answers to the question: What are the key points of the HALT process?

- Develop HALT with discrete and combined load cases and step stress conditions.
- Employ time compression to find design weaknesses faster.
- Ensure test coverage with internally designed built-in-test capability and the use of adequate external test equipment to detect faults and guarantee high test coverage.
- Understand the capability of the design.
- Define the fundamental limit of technology (i.e., determine the operating and destructive limits).
- Build confidence in the design performance through knowledge of the design margin and factors of safety.
- Precipitate and detect latent failures and defects (i.e., search for the weak links).
- Detect all failure sources (hardware and software).
- Increase stress conditions until a unit fails or is destroyed or to a predefined limit.
- Indicate sensitivity to stresses.
- Determine the product's robustness or improve the product's robustness.

- Correct inherent design and manufacturing process flaws.
- Conduct a failure analysis on every failure.
- Continue HALT through an iterative process (with corrective action, this improves product design margin).
- Determine the size and weight of the equipment to be tested to determine the chamber size and the weight-handling capability of the vibration table.
- Design test fixtures with high mechanical transmissibility and low thermal mass.
- Use HALT to develop highly accelerated stress screen (HASS) and highly accelerated stress audit (HASA) profiles for production testing.
- Establish groundwork for future reliability testing to generate reliability performance metrics.

To consider the meaning of HALT from another angle, let's analyze what HALT is not.

- HALT has no pass/fail criteria.
- HALT is not a qualification test.
- HALT is not intended to find and fix all failures. The failures that affect product reliability should be resolved and an awareness developed of failure modes at stress conditions well beyond the design specification limits.
- HALT is not used to calculate the demonstrated reliability of products. Reliability tests such as accelerated reliability testing or accelerated life testing are used to calculate reliability performance metrics such as mean time between failures.
- HALT is not a method to calculate acceleration factors for failure mechanisms.
- HALT does not always test to destruction limits if the limit is already well understood through vendor data or other analysis. HALT may use preestablished test limits (preset stress limits in the HALT plan).
- HALT does not require a long time, such as weeks or months, to find weaknesses in a product design and suggest fixes in a product design to improve reliability.

As an exercise, consider why HALT became widely accepted in the commercial product design marketplace in the early to mid-1990s.

HALT achieved early acceptance in commercial design applications due to its ability to rapidly identify design weaknesses early in the product development phase. The initial success of HALT resulted from short cycling the design process to determine the design margin and establish design maturity faster than in the traditional reliability test approaches. Short cycling compresses the time spent in development testing by revealing defects and product vulnerabilities faster than common accelerated test methods and their associated acceleration factors. During development testing the precipitation of latent design and manufacturing defects and early life wearout mechanisms are accelerated faster than in the normal evolutionary process of detecting failures in the field (customer use applications).

TIME COMPRESSION

One key point of HALT is time compression. Time compression in HALT does not result in acceleration factors that may be calculated for any particular stress condition, but does result in the acceleration of failure mechanisms. This time compression results from:

- Stresses that are applied higher than expected in the field application environment
- Stresses that are applied beyond the product design's specifications and operating requirements
- Higher vibration stresses beyond where Miner's criteria can approximate accumulated fatigue damage
- Hotter test conditions and colder test conditions beyond where thermal acceleration models exist to substantiate an acceleration factor for thermal stress conditions
- An increased environmental cycling rate that can be modeled by exponential acceleration of stress versus the number of stress cycles
- Higher-temperature cycling conditions beyond where Coffin–Manson's model is useful for calculating the acceleration factors from temperature-cycling stresses. Coffin–Manson's model is useful when the temperature-cycling stresses are constant and applied uniformly, but this is not the case for HALT because HALT uses accelerated temperature-cycling stress conditions.

Time compression or failure mode acceleration is accomplished through exercising a single stress condition or combinations of stress conditions. These conditions typically are some combination of temperature extremes (hot and cold), temperature cycling (or thermal shock), vibration (typically, three-axis random vibration with six degrees of freedom), mechanical shock, humidity, power, operating profiles, voltage, current, duty cycle, and frequency. Stresses may be applied individually or in various types of combinations, such as combined temperature cycling with voltage cycling and vibration cycling. For each stress condition, an acceleration factor may be calculated and measured. Accelerated life testing (ALT) is performed where one output from this type of test is an acceleration factor based on an acceleration model in terms of test time equated to actual customer use time. For example, ALT with a temperature-cycling slew rate of 10°C per minute may be found to have an acceleration factor of 10, which means that 1 hour of testing is equivalent to 10 hours of customer use in a normal field operating environment. If the ALT condition changes to thermal shock with a slew rate of 100°C per minute, the acceleration factor may be 100 and the value of the time compression increases by a factor of 10 compared to temperature cycling.

TEST COVERAGE

Another key point of HALT is test coverage, which is used to analyze the extent of the ability of the design and external test equipment to identify faults in the hardware being tested. Test coverage is another term for fault coverage or probability of fault detection. It is meant to reveal operational and environmental vulnerabilities. Test coverage is not a method to determine if all the design specification requirements are met by the product or system design, nor is it a test of the product or system to meet its requirements.

In HALT an electronics box, consisting of multiple circuit cards packaged within an enclosure, is exposed to the combined accelerated environments. During this testing, the electronic test points must be accessible to permit monitoring circuit performance parameters. This is done when a formal test station is not available. Test equipment probes are attached directly to test points and circuit traces or component leads to ensure that the design is operating as planned. The duration of each step within a step stress profile is based on the soak time and the length of time required to perform a test. The soak time is the amount of time needed to determine that the equipment under test has reached thermal equilibrium. After the soak time, the additional time in the step is the duration to perform the electrical test on the equipment under test. The amount of stress increase after each step is based on the sensitivity of the equipment to the stresses and how much time the analyst wants to take to determine the fundamental limits of the technology, where the operating design limits are for determining the operating and destruction margins.

Key Points of HALT Test Coverage Analysis

1. Must have good test coverage of all operating states, branches, and conditions for full detection. An example of good test coverage might be more than 90% of the circuit tested for faulty conditions. Without good test coverage, you don't know how effective your testing has been and how many faults might remain.

2. An undetectable problem will become a detectable problem at some point in a product's lifetime.
3. An undetectable problem can be an intermittent problem that will plague the product over its field deployment even when the test coverage is 100% of all functionality tested. Test coverage should include the ability of the test to detect time-dependent failure mechanisms, such as probabilistic mechanisms, as well as functionally dependent failure mechanisms.
4. Test coverage analysis is measured as a percentage of the design that is testable and able to detect faults, such as the product's built-in-test design, which has 97.5% probability of fault detection, and 99.8% probability of fault detection when connected to external test equipment.

ENVIRONMENTAL STRESSES OF HALT

A critical decision in HALT planning is the selection of HALT step stresses and load cases for various environmental stress conditions. Typical stress conditions in HALT are separated as load cases. Within each load case are step stresses. Stresses may be applied individually or in various types of combinations, such as a combined temperature cycling with voltage cycling and vibration cycling. A typical HALT plan may have six types of test profiles (load cases) with various step stresses. Typical load cases and step stress conditions are:

1. High-temperature step stress
2. Low-temperature step stress
3. Temperature-cycling step stress or thermal shock transitions (Figure 1)
4. Random vibration step stress
5. Combined environment temperature cycling and random vibration
6. Slow ramp detection screen

Any of these load case profiles could be executed with combined electrical operation conditions, power cycling, and input voltage cycling or voltage variation to further strengthen the test capability. Voltage cycling or voltage variation may include:

- On–off cycling
- Input ac or dc voltage static (i.e., margins)
- Input ac or dc voltage dynamics (i.e., dips, interruptions, and variations)
- Frequency margins (applicable mostly for transformer-coupled linearly regulated converters)

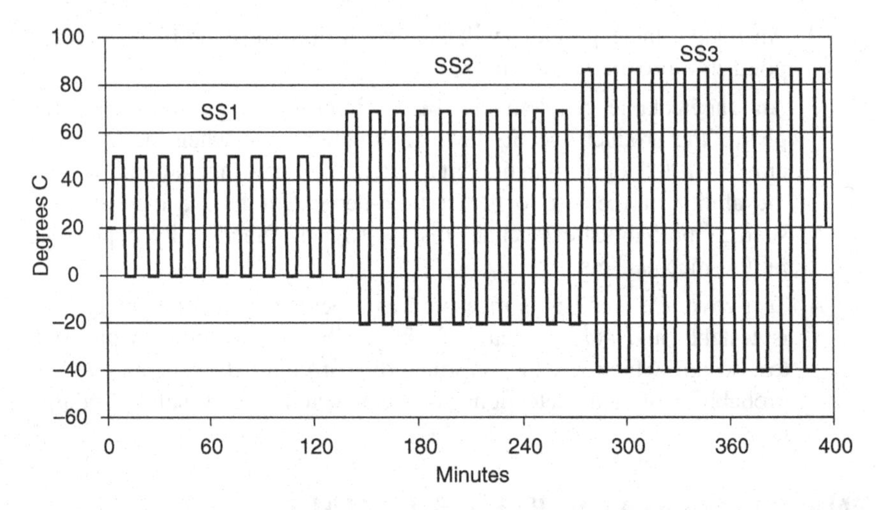

Figure 1 Temperature-cycle step stress load case.

An example of a temperature-cycle load case with three step stresses is shown in Figure 1.

SENSITIVITY TO STRESSES

A key advantage of HALT is the ability to determine product sensitivity to stress conditions. For any type of load case involving step stress conditions, the HALT engineer should decide how many steps and how large the incremental steps should be. Step stresses may start as large or small steps, depending on how important it is to determine the operating limit versus the destruction limit. The operating limit is the point in HALT where intermittent operation occurs but the product recovers at ambient test conditions. The destruction limit occurs at the stress level where the product cannot recover from a failure mode when operated at ambient levels.

For the example in Figure 1, you will notice that there are three steps between the temperature ranges 0 to 50°C and −40 to 90°C. Each temperature-cycle step stress increase is 40°C. The step increases could have been 10°C or 20°C if the goal of the HALT test was to identify the sensitivity of various failure modes to stresses at various stress levels on the way to the −40 to 90°C range limit. There is nothing scientific in the selection of this step increment other than the speed that it will take to reach the operating or destruction limit. The duration of each step is based on science, however. The step duration is the time it takes to fully exercise or test the function which may be sensitive to the stress. If the test takes 5 minutes to run at each step versus 1 hour at each step, this difference in test time at each dwell period could contribute to the decision to apply larger step increases instead of smaller steps during the load case. The test may be synchronized so that the electrical test profile is out of phase with the temperature cycles so that a certain electrical test step occurs at a different place in the thermal profile.

The HALT analyst determines the sensitivity to stresses by exercising the equipment in small steps in the stress area where operation becomes intermittent. As the test stress is increased in large steps with continuous test stimuli and detection, and the test records degraded performance, the analyst should bring the equipment back to a stress level where reliable operation is restored, and then start the step process again. This time, as the step approaches in the step stress sequence the point at which intermittent operation occurred initially, the stress should be increased in smaller steps, monitoring the performance continuously to determine if the equipment repeats the intermittent operation at the same stress area as recorded previously. This stress area is the condition in HALT where the product potentially has no design margin. If the failure is still intermittent (e.g., soft failure) at the same stress area, the operating limit is verified, and the test stress steps continue beyond this limit to the point where intermittent operation occurs more frequently and then leads to a hard failure condition in which equipment operation does not recover when the stress is reduced to specification levels or lower stress levels under nominal conditions. The destruction limit is the stress level where a hard failure occurs, repeatedly, and may be considered a design weakness that limits product robustness. Two or three product samples should be available for HALT to prove the case for a known destruction limit. At this limit, the fundamental limit of the technology is reached. If only one sample fails and another sample passes at the stress level, which was predicted to be the potential destruction limit, this may demonstrate the case for variability in the manufacturing quality or process of the parts or components supplied, or a potential design weakness.

Test stimulus may be applied in several ways. The test stimulus may be electrical or mechanical measurements recorded based on known test inputs or test signals. Test signals may be injected at specific nodes on the circuit, continuously stressing the circuit. Test meters are used to monitor the circuit outputs and detect failure conditions. Software test routines may be injected into the circuit interface and looped repeatedly to accelerate and precipitate the detection of latent faults, which may be caused by probabilistic failure mechanisms. Looping test routines constantly stressing the circuit functionality within a few hours accelerates the latent faults that would otherwise surface after many months or years of operation by the customer. The failure mechanisms may appear initially as intermittent probabilistic failures, but will increase in frequency of occurrence and become more repeatable with further accelerated testing over time until they become deterministic hard failures.

Time compression or failure mode acceleration is accomplished through exercising a single stress condition or combinations of stress conditions. These conditions result in an acceleration of failure modes, typically with combinations of environmental and electrical step stresses integrated in a single load case test profile, which may include temperature extremes (hot and cold step stresses), temperature cycling (or thermal shock), random vibration (typically, three-axis simultaneous repetitive shock random vibration with 6 degrees of freedom), mechanical shock (such as drop testing), humidity, power, operating profiles, voltage, current, duty cycle, and frequency.

DESIGN MARGIN

HALT provides a fast means to determine the product design margin between the specification and operating limits, and between the operating and destruction limits. The specification limits are the documented requirements, which highlight the capability of the design. The design margin between the specification limits and the operating limits, also called the *operating margin*, usually has very little overlap between the design strength and the applied stresses expected for the product. A small overlap between the design strength and the stresses applied translates into no physical fatigue, or physical fatigue that develops slowly over time and stress. This condition is depicted in Figure 2. The area outlined in heavy lines shows where the overlap occurs and reflects the operating area where failures are likely to occur. As the overlap between stress and strength increases, physical fatigue develops more quickly, and the operating area grows where failure probability is likely. Early life wearout mechanisms occur when the physical fatigue accumulates to the point of failure or damage. When the stresses surpass the strength of the design, there is negative design margin, or overstress conditions. In the overstress condition shown in Figure 3, the stress area overlaps the strength area entirely, resulting in a 50% probability that the product will fail when the product is operated. When the design strength becomes greater than the stresses applied, the positive design margin is increased.

For example, if a table was required to withstand weights of 100 lb dropped from a height of 1 foot, one might design the table to handle 200 lb from 1 foot. This increased design strength provides 100% design margin, which has a direct correlation on the high reliability or robustness of the table. The test measurements achieved from HALT output provides results in terms of the percentage of design margin for that product design, which is very useful to the design engineer.

The design margin for thermal performance of the product after completion of HALT may be determined to be 50%, as an example. This 50% design margin means that the design operating limits are 50%

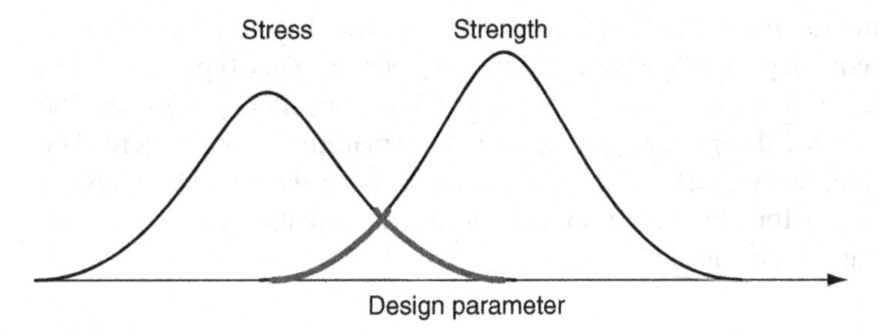

Figure 2 Stress-strength overlap.

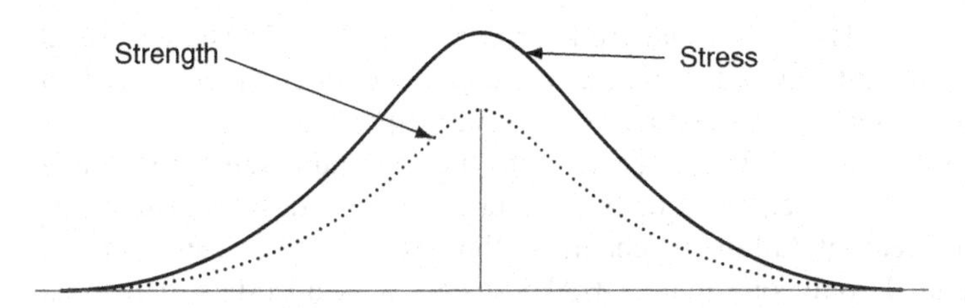

Figure 3 Example of overlap resulting in 50% probability of failure.

beyond the specification limits. For example, if the specification for a particular design states that the operating high-temperature requirement is 100°F, or 30°F above room temperature (70°F), a 50% design margin for high-temperature conditions means that the design will operate reliably up to 115°F (30°F × 0.5 = 15°F) before failures are highly probable.

Using the example above, for a design with 50% design margin, this means that the design is capable of operating at a high temperature that is 50% above the design specification's high-temperature operating requirement. The high-temperature operating limit is the actual level of operation that the design is able to withstand up to intermittent operation and the appearance of soft or hard failures. This level of design margin usually has some amount of overlap between the design strength and the stresses applied. When the design strength does not overlap and exceed the design stresses, no failures will occur. Some distribution of failure points may follow a normal distribution curve. The upper tails of the stress distribution may enter the lower tails of the design strength distribution in some units. When design strength distribution begins to overlap design stress distribution, the design experiences a nonzero probability of failure. As the strength and stress overlap continues, or increases, physical fatigue accumulates at an accelerated rate. Early life wearout mechanisms of a design are precipitated with prolonged stress exposure. When the design performs continuously at the operating limit, the design margin is approaching zero.

The destruction limits are the levels of stress in which intermittent failures increase in frequency until the design fails to operate. This type of failure, which is not able to recover, is called a *hard failure* or *patent failure*. At the destruction limit, the design margin is zero and the margin is close to zero. The accumulated fatigue stresses exceed the strength of the design. The physical stress is too much for the design strength to handle. At this level, materials fail (i.e., fracture, melt, or vaporize).

SAMPLE SIZE

The sample size for HALT is usually one or two units. This decision is based on the cost of the units and knowing that the units will not complete HALT in a condition to be sold to customers. When given a preference, I select three to six samples for HALT. Three samples are the minimum sample size and are initially tested in a sequence or in parallel to identify design weaknesses and early life wearout failure mechanisms that isolate a root-cause failure pattern or trend. If only one unit is HALT tested, it is uncertain if a trend has been detected. If two units exhibit the same failure mode and root-cause mechanism, there is a higher statistical significance that a trend has surfaced compared to a single failure. If three units are tested, this will increase the statistical significance of a trend if all three detect the same failure. It will also result in identification of a trend if two out of three units fail for the same reason.

Once the root-cause resolution is determined and the corrective action implemented to the design, another sample of three units is selected. These may be units that are reworked or repaired from the original test sample population, or it may be three new units built with the design change incorporated at initial assembly production. This depends on how many samples were build for the test phase, on whether there are more units available to conduct a retest of the HALT, and/or on the extent of the design change corrective action.

CONCLUSIONS

With repeated HALT, using appropriate environmental and electrical step stress conditions, load cases, and a sufficient number of test samples, HALT is useful to improve product reliability and to assure adequate design margin within customer application environments and stress conditions. This is true only if failures are identified and corrective actions are taken during and after HALT to increase the design margin. Reliability improvements can be made when the design is changed to either increase the design strength or lower the design stress, or both.

Acknowledgment

This chapter was developed from a four-part series of articles on highly accelerated life testing published in four consecutive IEEE Reliability Society newsletters between 2008 and 2009.

REFERENCE

[1] Hobbs, G. K., *Accelerated Reliability Engineering*: *HALT and HASS*, Wiley, Hoboken, NJ, 2001.

Chapter 12

Design for Extreme Environments

Steven S. Austin

OVERVIEW

In a nonstandard environment, a normal application used for engineering, safety, reliability, and testing requires materials and testing using criteria outside the parameter of standards of acceptance. Harsh environments lead to cracks, higher failure rates, and destruction of personnel and material with a higher risk of loss. Planning and designing for an extreme environment incorporates new ideas with proven materials and lessons learned. This approach uses a subjective method of design and testing for all various types of extreme environments.

DESIGNING FOR EXTREME ENVIRONMENTS

Human beings have fought the elements since the beginning of time, which from necessity of survival led to wearing furs, wool, cotton, nylon, polyester, and now Gortex and Thinsulate. Metals, plastics, lubricants, and glass have all taken on new identity ratings when used in extreme environments. To give credit to the pioneers of designing for extreme environments, the U.S. Army and National Aeronautics and Space Administration (NASA) are leaders of technological advances in this area. Ruins of World War II vehicles and equipment are still visible today in such countries as Morocco, Libya, Syria, Italy, and Tunisia. Engineers have marveled at Rommel's and Montgomery's

Design for Reliability, First Edition. Edited by Dev Raheja, Louis J. Gullo.

battles in Tunisia, but the fact remains that in the desert heat, sand, and low humidity, the moving parts of all vehicles are destroyed while unused metals are preserved. Also, along the Alaska Highway, built in 1942 from Montana to Delta Junction, Alaska, equipment used by the engineering company can still be seen. The workers who spent the year building the highway faced extremes they were totally unprepared for: environments that no one had tried to conquer before. They improvised on clothing, equipment, and fuels to keep themselves alive and their equipment running. The extremes that engineers and soldiers faced laid the groundwork for an Army testing and training center at Delta Junction, due to the region's extreme cold temperatures during the winter.

DESIGNING FOR COLD

Molecules that are human-made have to be both pliable and rigid to withstand rapid changes in weather. Companies contracted to build materials that meet government tolerances and specifications for normal conditions may lack the ability to meet extremely cold condition and have to collect data from environments not well known to us. An example is plastic, used widely in everyday life. Testing under normal conditions has proven that plastic has many versatile uses, however; plastic becomes molecularly unstable in extreme cold. From the 1940s through the 1970s, plastic use was minimal; steel and aluminum were the standard materials. Computers have brought about new technology, and scientists are developing new and futuristic materials. The reliability of human-made materials is being proven either through consistent use or standardized testing. Sample testing is one method used by large companies. Quality standards set by the American Society for Quality (ASQ) are widely used. With the advent of six sigma and lean manufacturing, quality and reliability are providing calculated numerical results. Management and engineering are standing up and taking note of the cost of manufacturing and of high-quality products with fewer defects. This principle also applies to materials and equipment produced to be used under extremely cold temperatures. When life or life support is threatened, reliability becomes the greatest common factor. Community dwellers in extremely cold environments understand how depending on life-sustaining equipment equals existence. Heat transfer equates to comfort, food, energy, entertainment, and transportation. Materials used to construct living facilities, factories, and businesses have resistance factors designed to lower energy loss.

Planning a design for cold requires consideration of many environmental factors, such as moisture, static electricity, freezing, seismic activity, and wind. Scientists continue to study the changing environment and attempt to predict warming trends for cold regions of the Earth; however, evolution of the Earth is measured geologically through the soil strata layers. Soil or rock forms the foundation for all structures. Given that the Earth's layers move during earthquakes and volcanic eruptions, structural load testing in simulators and real environments can predict failure points. Using failure modes and effects analysis (FMEA) evaluation of inside structural components can lead to the identification of a single-point failure on paper. A nondestructive evaluation or test can determine cracks or breaks in a facility's structural integrity.

With major populations moving out of cities to suburbs, utilities and services have to be extended to match the expansion. Usually, jobs or government amenities drive the nomadic movement of people to push to the outer fringes to live. Since Thomas Alva Edison's invention of the light bulb, people have harnessed natural gas, oil, water, sunlight, wind and nuclear energy to sustain the lifestyle of the twentieth and twenty-first centuries. This marked use of energy has a direct economic and metric connection to reliability. In third-world countries, extreme conditions are simply dealt with, whereas in prospering countries, maintaining energy reliability has become a way of life. In extreme cold, heat is the predominant requirement. Keeping food from getting too cold is an issue. Stand-alone containers designed to withstand extreme cold and do not require energy to maintain a constant temperature have been designed but are too costly for the common household. Use of common materials sold in the marketplace remains the predominant supply for construction. When designing for extreme locations, factors such as use, protection required, gradient temperatures, wind loads, seismic activity, and water flow must be planned and constructed to higher standards than the U.S. Army Corps of Engineers' military specifications. Major strides in plastics and ceramics are continue but require proof testing before mass implementation.

The effects of extreme heat and cold were tested after the Korean War as directed by President Dwight D. Eisenhower. He saw the devastating effects of cold on troops and equipment in Korea and in Europe. Clothing for soldiers was not designed for extreme conditions, and equipment failures with unserviceable weapons and engines made it difficult to fight battles. The Cold Regions Test Division, now known as the Cold Regions Test Center (CRTC), at Delta Junction, Alaska services the entire Department of Defense and other agencies in the testing of uniform and uniform items, explosives, materials, vehicles, and weapons. Many tests at CRTC are conducted in extreme conditions of -50°C.

How humans work in these harsh conditions on a daily basis is a feat in itself. Properly layered clothing, including a warm cap, a facemask, goggles, moon boots or -100°F-rated boots, over-pants, and a parka, are the standard wear under cold conditions (Figure 1). Commercial vehicles have to be supplemented with electric heating pads on engine blocks, oil pans, starters, batteries, and water pumps. Without these, engine fluids gel and eventually freeze solid, at which point heat has to be applied to thaw or regenerate them into a liquid state. New designs for synthetic lubricants are being tested and

Figure 1 Dressing for extreme cold. (Photo courtesy of Mike Kingston, Cold Regions Test Center, Alaska.)

used in this type of environment. Plastics also have problems, as they become brittle and shatter like glass.

Designing for this environment requires thinking and designing outside the box. Cashing in on this are engineers at most manufacturing plants worldwide. Designing in a bubble with perfect conditions will not and cannot work in the harsh extreme conditions of the Alyeska pipeline.

DESIGNING FOR HEAT

Operations Desert Storm Iraqi Freedom have validated U.S. Army test data derived from the Aberdeen Proving Grounds and from the Yuma Proving Grounds in Arizona which indicated that desert-proofing equipment before deployment adds to its operational life. The military is only one recipient of these test results; contractors and government workers who build military equipment gain valuable insight into generating high reliability at lower cost with proven data. What do engineers look for in extreme designs, and how can structural, mechanical, and electrical components be made to withstand sand and heat to operate in these extremes?

An example of extreme heat is the eruption of Mount Saint Helens in Washington, where 57 people died during 9 hours of continuous eruption. Given the energy release of a 10-kiloton nuclear bomb, what could be built to withstand the heat, blast overpressure, and some 3700 million cubic yards (4847 million cubic meters) of blown material? This did not count the millions of trees and mud slides that washed down, leading to the destruction of the valleys and rivers below. The eruption of Mount Saint Helens was triggered by an earthquake, which set off landslides on the north face of the mountain.

The southern California fires in 2003 are another example of designing for heat. In a matter of hours in San Diego County, California, around October 2003, a total of 2500 homes were destroyed. Sixteen people were killed and over 500,000 acres (200,000 hectares) were burned to an unrecognizable moon landscape with a few ruins left as mementos. What types of structures survived these fires? Concrete and adobe with little or no wood construction were in the lead, with steel and metal structures coming in a close second. Designers of homes and commercial structures who took the time to study the area's fire history found that architects who involved fire department planners and landscape artists with fire protection in mind survived at a better rate than did their nonplanning neighbors. Homes constructed in rugged landscape had almost a zero survivability rate over open-area and separated homesteads. Planning an emergency water supply is not always feasible, but those who did so were very proud when firefighters hooked up to civil emergency or private emergency water sources to fight the high-wind-driven fires.

Designing for extremes requires research of the location in which a person or machine will operate. Once data are collected, their ability to match the extreme must be validated. Given the data, manufacturers must select known material proven to withstand the extreme conditions (Figure 2), and the material has to be procured. Before release for final manufacturing, testing should be performed in a heat chamber such as an autoclave $+273^\circ$F or in a cold chamber or outside at -50°F to ensure that materials can withstand the environmental conditions. Once this passes the test, a test item should be produced and taken to the extreme environment for testing. If the test vehicle or item functions as designed, low- or full-rate production can be implemented.

Many more factors are considered in this process; however, testing before production is validation before investment. Reliability must be checked once a vehicle or item is in the field and in operation. Failure rates due to hot or cold stress fracture, seizing parts, failure to perform, corroded or separated connections or wiring, adhesion, lubricant, and related issues must be documented and included in operations research methodology for incorporation into new

Figure 2 The fate of equipment under extreme conditions. (Photo courtesy of Mike Kingston, Cold Regions Test Center, Alaska.)

product or modified product lines. Changes or modifications in design are generally extremely costly and require major engineering redesign; however, working in extreme environments generates issues of its own.

Humans can perform for only limited amounts of time under extreme conditions. Therefore, testing under such conditions involves constraints on time, energy, money, manpower, and in some cases, data collection and dissemination. Engineering feats have been conquered all around the globe with noteworthy success. The Chunnel from France to England is a marvel of mining, civil, mechanical, and electrical engineering disciplines combined.

NASA has been testing for extreme environment conditions since the 1960s and has played a major role in metallurgy development for conditions of cold regions. Lessons learned from space are a valuable resource, as the Moon's surface is known and how to exist on it comfortably has been proven. Building in jungles such as the Panama Canal has also been a major engineering challenge. To build similar capabilities today with the machinery available would take a fraction of the time, as proven by the Tennessee–Tombigbee human-made waterway. Nuclear plants are an alternative to growing energy source reliability, and the radiation environment required to sustain and monitor reliability is just another case of extreme environments. With cost and production being the force driving today's market, test budgeting is falling behind and prototype ventures appear to be leading the way. Without ensuring that a product can withstand the environment, companies will lose, and some losses will be unacceptable (e.g., the *Titanic*).

Safety, reliability, and testing are only a few of the criteria established to lead the way to better product enhancement, but the market will drive the requirements for supply and demand. Extreme environments such as alternative fuel sources for transportation in today's world will drive engineers to design vehicles that are not powered by fossil fuels or electricity. However, humans have developed mass systems such as trains that can function in extreme conditions such as snow and ice or blistering heat. Reliability can be high and cost low when engineers operate in a free market without stifling regulations.

Many opinions have been formed about exploring for natural gas in the northern regions of Canada and Alaska (Figure 3). With today's data and technology, a process that took seven years in the case of Alyeska pipeline can be done in much less time with a much greater safety factor than imagined. Providing products through engineering with regulatory guidance is the art of balancing need with supply. Given the means of product failure research, analysis, and modeling to establish a high-quality product at the lowest price for the environment using the least natural resources meets the goal of this chapter. A quote by Wilbert E. Scheer states it best: "Tolerance is the oil which takes the friction out of life" [1].

Designing in a safety factor at an affordable cost not only ensures the stability and longevity of a product but provides a level of confidence from the user or buyer that the product will not fail, damaging property or injuring someone. The normal safety factor based on stress analysis or product testing is not always the margin of safety when dealing with extremes. Not only

Figure 3 Terrain in the far north. (Photo courtesy of Alaska Wildlife Fish and Game Division.)

do extremes change the properties of human-made materials but they cause materials to fail just as humans fail in extremes when not protected properly. Take, for example, the fact that the safety factor of plastic when in a normal climatic environment may give a 4 : 1 yield safety margin, but in an extreme cold environment the same material may have only a 2 : 1 yield and fail much faster under the same test pressures. An ice road trucker weighs the risk and, based on core ice samples, determines if the ice is strong enough to carry a load across frozen water, based on temperature, thickness, and total ground-bearing pressure weight. Do these truckers have mathematical formulas to derive a calculated risk, or do they take chances based on past experience? The pressure to deliver is for the individual company or operator to assess.

Research and development, manufacturing, and sales personnel depend on the ability to establish and maintain the credibility of a product to perform under all conditions. The military uses 1×10^{-6} as the safety risk factor, and this does not guarantee that an item will not fail, but it affords a greater degree of confidence that 1 in 100 million could fail. Therefore, with extremes, any error should be made on the side of caution, using a 1×1^{-6} safety factor [2]. With this number in mind, reliability testing can provide the validity of this number in a given product, but there is a cost associated with the testing. Stringent product testing in Japan and Europe has become the norm, not the exception, whereas in the United States, product testing has been performed on a sample or random basis.

Without a safety rating attached to a product, its confidence for sustainability may be taken for granted. Safety, quality, and reliability are disciplines that predict and monitor a product's performance. Extreme conditions demand these disciplines, as they are the final lines of fault detection and thus of product safety. Keeping lines of communication open between engineering and safety personnel allows for a sharing of data that must be captured and built into the system as lessons learned, to prevent failures.

Revisiting lessons learned from the ancient Romans, Greeks, and Egyptians, aquaducts, engineered stone pillars, statues, and arena remain standing from 2000+ years ago to remind us that no matter what the extremes, products can be built to withstand extremes over time. Historians have gathered a wealth of information when examined from a system reliability standpoint, leading to clues as to the resiliency of these artifacts to withstand extreme cold, heat, lightning, wind, earthquakes, volcanic eruptions, and the most destructive of all natural forces—water. A classic example of a recently failed extreme design was the seawall levee protecting New Orleans, Louisiana. The original design lasted for 60+ years, but was not properly maintained and repaired thus weakening the structure. When the levee broke, the results were catastrophic.

Prior to the European industrial revolution, hand craftsmanship was the main manufacturing process. Now computers and robots perform many of the tasks once performed by workers on assembly lines to a perfection tolerance only dreamed of by our ancestors. With this knowledge obtained and the future of the workforce being a mixture of humans and artificially intelligent computer-driven machines, the only failure point in the process is materials when it comes to designing for extremes in the future. Since the 1940s, exotic metallurgy has shaped the design of aircraft, tools, and machines. Uses for precious metals have soared and the development of substitutes has stagnated. NASA research scientists in joint ventures with scientists from around the world have designed space technology to meet ultimate extremes, such as freezing cold, asteroid impacts, and zero gravity. Future generations will have the speed of constantly upgraded computer technology to retrieve, recalculate, and reinvent new materials and processes to be shaped into products for extreme designs such as bridges, tunnels, and transport systems. Designing for extremes is a constantly evolving process and must continue to be undertaken seriously to carry us through future generations.

REFERENCES

[1] http://www.best-quotes-poems.com/tolerance-quotes.html, or http://acquirewisdom.blogspot.com/2007/08/learn-to-tolerate-one-another.html.

[2] *System Safety Program Requirements*, MIL-STD-882C, U.S. Department of Defense, Washington, DC, Jan. 1993.

Chapter 13

Design for Trustworthiness

Lawrence Bernstein and C. M. Yuhas

INTRODUCTION

Trustworthy means that a software product or component is safe, reliable, and secure. Software design is the important first, not the final, step toward creating trustworthy systems [1]. In this chapter we provide design techniques and constraints on the software implementation that will lead toward that end. The goal of the design process is to create simple and concise solutions. Simplicity improves reliability, and conciseness reduces the time and cost of implementation [2].

Often, software system development is dominated by schedules and cost. Sometimes performance and functional technical requirements become an issue. Rarely has trustworthiness been considered in any but the most critical systems, but this is changing. Society as a whole is beginning to recognize that not only must software designers consider how the software will perform, they must account for the consequences of failures. Trustworthiness encompasses this concern.

Software *fault tolerance* is at the heart of building trustworthy software, although that may seem a contradiction in terms. Trustworthy software is stable. Therefore, it must be sufficiently fault tolerant that it does not crash at minor flaws and will shut down in an orderly way in the face of major trauma. Trustworthy software does what it is supposed to do and can repeat that action time after time, always producing the same kind of output from the same kind of input. The National Institute of Standards and Technology (NIST) defines *trustworthiness* as "software that can and must be trusted to work dependably in some critical function, and failure to do so may have catastrophic results,

Design for Reliability, First Edition. Edited by Dev Raheja, Louis J. Gullo.

such as serious injury, lost of life or property, business failure or breach of security" [3].

Software execution is very sensitive to initial conditions and the external data driving it. What appear to be random failures are actually repeatable. The problem in finding and fixing these problems before a design is released, or even if a problem emerges once the software is in use, is the difficulty of doing the detective work needed to discover first, the particular conditions, and second, the data sequences that trigger the fault to become a failure. The longer a software system runs, the more likely it becomes that a fault will be executed and precipitate a failure. Reliability can be improved by several techniques, such as the following [4]:

Limiting the execution time,

Making sure that data are within specified bounds,

Investing in tools,

Simplifying the design,

Doing more inspections or testing in development, or

A combination of all of these factors.

Trustworthiness has the implied quality of "no surprises." Users (stakeholders, including the end user and the system integrator) have good reason to expect predictable behavior of the software under all anticipated ranges of inputs and environments. This raises questions about how well the operational context and likely evolution of the system are understood. It also implies that to be trustworthy, software has to be robust in the face of unexpected uses and evolutionary change, or at the least, if it breaks, it will fail in a tolerable manner. The software engineer must design for transparency and for better characterization of robustness in the operational environment expected [5]. For example, an air traffic control system that occasionally just stops working is clearly less trustworthy than one that stays alive. If another stays alive but occasionally displays hazardously misleading information, is it more or less trustworthy than the one that crashes? The first is unreliable and the second is unsafe. Both are untrustworthy.

Here is a checklist that you can use to get a sense of the trustworthiness of software:

1. Does the team have adequate domain knowledge and experience?
2. Are requirements and features completely and clearly defined in the context of the problem domain?
3. Do the architecture and design solve the problem adequately, and is the software execution reasonably bounded?
4. Is the software technology mature, well supported, and understood?

5. Is the architecture appropriately safe, reliable, and secure?
6. Were steps taken to assure trustworthiness? Was refactoring used?

Software Structure Influences Reliability

David Parnas, a great software engineer, uses *structure* to mean both how a software system is divided into modules and also the assumptions that the various modules make about each other [6]. A module is intended to be a unit of work assigned to a single programmer or a group of programmers. If the module is also a component, it can be constructed with no knowledge of the internal structure of other modules. The combination of structures, rules of composition, and interface design comprises the architecture of a software system.

Parnas explains that reliable software need not be correct software: "We may consider the system reliable in spite of faults if either (1) the programming errors do not make the system unusable (e.g., format of output, erroneous output that is easily detected and corrected by the user) or (2) the situations in which the errors have an effect do not occur very often, and the situations are not especially likely to occur at moments when the need for the system is very great" [7]. When software execution is limited appropriately, a latent fault will not be triggered to induce a failure.

Unreliability often results when modules are designed on the assumption that nothing will go wrong. In this case, the software structure naively assumes that everything outside the software module behaves correctly. Alternatively, the structure can incorrectly assume an "all or nothing" approach with no definition of degrees of imperfect behavior. These precautions must be taken to avoid such problems:

1. Specifications for the system and each of the modules must define the desired behavior and provide tolerable alternatives when perfect behavior is not obtainable.
2. Interfaces have considerable trouble potential. Both those between the system and its environment and those between modules require examining module behavior in handling off-nominal conditions. Explicit specification is necessary not only for what the interfacing elements should do in normal cases, but also for handling likely off-nominal conditions. Module behavior should additionally be verified by run-time checks. When an error is detected, the required mitigating actions must be specified.
3. Include steps to inform affected modules about things that have gone wrong elsewhere in the system.

Table 1 Parnas Design Checklist for Reliability

Concern	Event	Does the architecture define:
Interfaces	Communications equipment failures during transmission or operator error	**1.** How the system will be informed? **2.** What information will be supplied? **3.** How the system should respond? **4.** What action the system should take if operators input messages with a priority beyond their privileges? **5.** How operators can indicate that they made an error?
Are message logs kept?	Failure of storage or storage area network	**1.** What corrective action should be taken when a log is lost? **2.** How fast messages must be recovered from any backup?
Fault tolerance	Memory problem	**1.** Whether the memory allocation module knows when a section of memory is malfunctioning and should not be allocated? **2.** The response to a detected deadlock? **3.** When a program erroneously asks for resources without releasing them, how other modules should be informed and how should they react? **4.** How memory leaks are detected and prevented?

Table 1 shows a possible checklist. The list of questions ends when the risk exposure is acceptable.

MODULES AND COMPONENTS

The designer translates customer requirements into system features and then into software functions. The software architect allocates software functions to modules that are relatively self-contained. The product of this effort is a logical block diagram. Each block contains a set of functions that are logically grouped into modules. Each module can, in turn, be subdivided into smaller logical modules. Modules may then be grouped into subsystems. The modules hide details of their internal structure from one another, but the features, performance, and trustworthiness of each module need to be understood.

Each module needs a design specification, a schedule for delivery to test and integration, staff hour estimates, and performance criteria. Programming techniques that optimize the use of electrical energy are highly desirable, and soon power usage budgets for software modules will be common [8]. Modules may be objects consisting of data structure and procedures or methods with states and defined behavior. The modules are tracked throughout the development process. When incremental programming is used or when there are multiple releases, a module vs. release matrix is useful for planning and tracking. It is helpful to have a planned release schedule so that module changes can be tested adequately and integrated more seamlessly into the design. The planned cycles also provide better traceability of design changes.

A component is a module that is independent, with clearly defined, small interfaces using a highly expressive interface language. Its implementation details are hidden from other modules, and data structures are accessible only through the interface language. A component follows a standard. If two components need to transfer data directly for performance reasons, they need no longer be separate and may beneficially be integrated into a single component. Components may evolve without requiring a complete set of system tests, but module changes typically require exhaustive release testing.

Interfaces

A single entry point to and a single exit from a module make updating interfaces more reliable. The module starts with a text preface explaining its functions, inputs, outputs, and a record of changes. A jump table follows to gain access within the module. All control and all data are passed to the jump table, which then transfers execution to the appropriate place in the module dictated by the input parameter. If this is too expensive in terms of execution time, direct transfers and passing data through memory are possible, but this obscures the dependency and makes it more difficult to reduce coupling. The goal of the trustworthy design is to achieve high cohesion within a module and a low incidence of coupling between modules.

Modules can be connected together and may have interchangeable parts, but they must be able to be designed, implemented, and tested independently. Low-coupled design adds a level of indirection, passing data and control parameters through a special interface object class. This makes the system more difficult to understand but easier to evolve. Parnas points out that the act of creating a system design that satisfies specific system objectives can be the assembly of modules, each purposed individually, in a unique configuration encapsulated to prevent unintended interactions.

Consider the problem caused when using a method that runs different paths, depending on the type of parameter, as shown in the following example:

```
Set value (String Name, Value) {
    If Name.equals ("height")
        Height = value
    if Name.equals ("weight")
        Weight = value
    if Name.equals ("Anything else")
        Error ("Input should be weight or height")
```

If the units change from English to metric, all modules will have to change. A better arrangement would be this:

```
Set height(integer, argument)
    Height = argument
Set weight(floating point, argument)
    Weight = argument
```

This example shows two valid ways for recording a person's height and weight. The engineering choice of which to use is based on the need to understand the code easily or, rather, to be able to make future changes easily. In the first instance the height and weight are stored as an arbitrary string of characters so that it is quickly stored but difficult to use in computations. The second method will result in computations without conversions from arbitrary strings to formatted numbers, making calculations with specified precision, simpler and faster. The software engineer must consider the anticipated use and life of the product when deciding which data structure to use. In any event, decoupling avoids the cascading problem, which makes it so difficult to change modules and their interfaces.

A strong module interface language makes it easy to construct designs that resolve problems identified in a Parnas design checklist for reliabity. The interface procedures may share data structures and layers of software services. The Flexible Computer Interface Form and TL1 are examples of such interface languages [9].

The principles of modular design separate solution details with horizontal decomposition so that different parts of the problem are handled by different modules. The interface design provides vertical decomposition as modules call one another. Some key ideas are as follows:

1. *Abstraction* leaves things that are unlikely to change *in* the interface, but implementation details that are likely to change are left *out* of the interface.

2. *Information hiding* shields design decisions that are likely to change from other modules. Each module's implementation is a "secret" to avoid dependencies on potentially changeable design decisions.
3. *Small interface languages* should be used to create consistent interfaces across a large system.
4. *The open and closed principle* identifies software components that are open for extension and repair but closed for modification [10]. A component is open if it is still available for extension by adding new code. For example, it should be possible to add new functions or fields to the data structures it contains. A module is closed if it is available for use by other modules, and it has a well-defined, stable description in addition to an interface that follows the principles of information hiding. The source code of such a module is inviolate. The only source code changes permitted are those needed to fix a problem, and even these must be controlled very carefully.
5. The *Liskov substitution principle* [11] states that pointers or references to an object class should be able to use objects of derived classes without knowing it. The importance of this principle is illustrated when one considers the consequences of violating it. If there is a function that does not conform to the Liskov substitution principle, that function uses a pointer or reference to an object model called a base class but must know about all the derivatives of that base class. Such a function violates the open/closed principle because it must be modified whenever a new derivative of the base class is created and leads to cascading changes.

Comparing Object-Oriented Programming with Componentry

The key idea in object-oriented programming is that software should be written according to a mental model of the actual or imagined objects that it represents. It attempts to create "verbs" and "nouns" which can be used in intuitive ways, ideally by end users as well as by software developers.

Software componentry, by contrast, makes no such assumptions and instead states that software should be developed by gluing prefabricated components together. This would create a "compile, integrate and test" software process akin to "assemble, wire, and test" hardware factories. The definitions of useful components can, unlike objects, be counterintuitive. It takes significant effort and awareness to write a software component that is easily reusable, as can readily be seen in this checklist:

1. Does a reusable component already exist?

2. Does it send and receive data only through formal and normalized interfaces?
3. Is it fully documented?
4. Is the range of its performance explicitly defined and budgeted?
5. Has it been tested for its functions and reliability, and has it been put under stress?
6. Are the test plans and results available?
7. Does it do robust input and output validity checking?
8. Does it pass back useful error messages?
9. Does it anticipate unforeseen uses?
10. Is building the component financially viable?

Note: Ariane 5, an expendable launch system, is a good example of software in which reuse did not work. The stress of the new application exceeded that of its predecessor.

POLITICS OF REUSE

A good idea lives or dies not solely on the basis of its merit, but also by a number of other factors that surround it. Can the creative culture of programming accept the idea of standardized parts, the business environment in which it is practiced, and the mathematical theory that supports it? Executives want to be competitive and tend to think that it is a matter of discipline in the programming ranks. Engineers, although historically fond of standardized parts, know that reuse without sound theory, excellent documentation, and control leads to disaster. The productivity gains promised by reusing software are problematic. Everyone is willing to entertain the concept, but the application and rewards remain stubbornly elusive. The potential of reusing some critical code containing errors discourages software engineers from adopting a reuse philosophy.

Many trends, however, suggest that reuse will become commonplace, even though large-scale reuse of software modules is difficult today. There are some successes with reuse to suggest that some projects are getting shorter development intervals at lower cost. Projects that use LINUX, UNIX, and its C and C++ libraries get 20% reuse without extra effort. Projects that adopt free and open source development are getting even more [12]. We project that by 2015 the software industry will have huge libraries of reusable components and split into small shops creating new components and huge shops integrating components into applications [13].

One big obstacle is that software revision is very difficult without the originator's help because so much code is indecipherable and takes too much time to decode. For example, there are long-reaching effects in object-oriented code because the early decisions in bottom-up design demand greater insight to isolate than those in top-down design. Managers do not value and praise their product's internal clarity; yet only clear code can be modified.

Data collected on the reuse of 2954 modules of National Aeronautics and Space Administration programs clearly demands the shocking conclusion that to reap the benefits of the extra original effort to make a module reusable, it must be reused essentially unchanged [14]. No change costs 5%; the slightest change drives the cost up to 60%. The most failure-prone changes are the small changes.

In the category of currently intractable problems is the impossibility of systematically reusing software across application domains. There is ongoing work in modeling application domains to capture the relationship between requirements and object types so that selecting these features can reuse software architectures. Also, reuse even in the same application domain is successful only when throughput and response time are not overriding concerns. It is not yet possible, but soon will be, to maintain an asset base of software modules except when they are in packaged libraries and when they are utility functions.

Where reuse is successful, there is high level of management attention to detail and a willingness to invest in the threefold increase in cost to design for reuse over single custom use. Software configuration management assumes that there is an existing base of software components from which the components of a specific system are chosen, complied, tested, and distributed to a user. Even then, exhaustive retesting is still required to root out what Jackson called "undesired interactions."

Advancing new technology is dicey. It is difficult to shift to new technology when a library of reusable modules exists. If you doubt that software is still a cottage industry, just count the number of software companies that have not advanced to level 3 of the Software Engineering Institute's Capability Maturity Model. There is considerable risk in being on the leading edge, but vast rewards are available to those who meet the challenge.

DESIGN PRINCIPLES

Poor design introduces unnecessary dependencies and makes components more difficult both to understand and to integrate into products. Good design is essential to eliminating barriers to reuse. Good reuse design requires good software design, using well-known software design principles.

Strong Cohesion

When code is cohesive, each code module contains a well-defined set of related functions and does not contain extraneous functions. Strong cohesion reduces the chance that a reuse component will contain features that are useless or unwanted by other products. For example, a component that mixes database access statements with data manipulation functions will not be attractive to a product that uses a different database or no database at all. Even worse, if the data manipulation algorithms in this component depends on the specific type of database access, the component will be unusable in products that store data in different ways. Highly cohesive code would have isolated database access from data manipulation and cleanly separated the two types of code into two modules. Following the practice of strong cohesion will help you to keep invariant features, such as data manipulation, separate from variant features, such as the type of data storage that a product uses.

Weak Coupling

Strong cohesion and weak coupling usually go hand in hand. Weak coupling between modules promotes the independence of each module, so that one module does not require knowledge of another. Weak coupling keeps a reuse component self-contained and more portable, at both compile time and runtime. For example, highly coupled code is likely to bring extraneous *include files* and functions into the compile; it is likely to produce overly large runtime components that will port poorly. High coupling usually means unnecessary and unwanted features, more complex product customization, and a greater possibility of disruptive side effects.

Information Hiding

Information hiding and the use of abstract data types allow a product to access a reuse component without needing to know how the component represents and handles its internal data. Information hiding is the principle that the internal data and operations of a module should remain hidden from other modules. Modules should communicate with each other only through clearly defined interfaces, and only communicate what external modules need to know. Information hiding promotes a clear external interface and enhances the ease of use of the component.

Inheritance

Inheritance is the ability of an object to include characteristics of a parent, or class, object in an object instance. Inheritance lets developers separate general and common characteristics from specialized and variable characteristics. Using the principle of inheritance, a developer can place commonalities in class objects and let product developers tune individual object subclasses to their needs. Too much of a good thing can be dangerous, as when too much inheritance leads to tightly coupled—therefore unreliable—software systems.

Generalization or Abstraction

To generalize or abstract a component is to ensure that the component will handle general, rather than specific, conditions. For example, if a list algorithm has been written specifically for byte data, the algorithm could be generalized to handle short, long, or double data as well. In some cases the algorithm could be generalized even further to handle any type of data. Generalization is the process of removing unnecessary detail and nonessential restrictions from a component. The decision to generalize a component is limited by the effort required to reengineer the component and by the range of variability shown in the domain analysis. If the domain analysis indicates that current and future products will use single key data access, it may not make sense to generalize database functions to handle multiple keys.

Separation of Concerns

This is the principle of organizing software so that you need handle only one issue at a time. Separation of concerns is what distinguishes a reuse component from a single-use, or product-specific, component. Separating concerns will dictate the design of the component by showing what can be fixed in the component and what must be customized.

Removal of Context

Similar to generalization, this principle applies to the component interface rather than the component internals. Context, in this sense, refers to the assumptions that the component makes about what is outside it: services, resources, infrastructure, and hardware platform. Infrastructure includes the operation system and available communication mechanisms. Context also includes assumptions about surrounding components and the messages, or information, they will pass.

An effective way to remove context is to design weak preconditions into a component's interface. Preconditions are expectations that a component, or interface function, has about the data passed to it or its state when it is called. For example, expecting that an input parameter will always contain valid data is a strong precondition. This precondition requires external components to check the validity of the parameter before they pass it, and may make it difficult to integrate this component into various products. To weaken this precondition, the component should verify the validity of the input data itself before continuing.

Expecting a particular order of events is also a strong precondition. For example, a component may assume that an external component will call its initialization function before any other functions, or a passive component may assume that its caller will call entrance and exit functions for it. To lessen such preconditions, the component can store information about its current state in global flags. For example, an initialization function could set an *INIT* flag that would be checked by all subsequent functions that require initialization.

Removal of context can be onerous. The domain analysis determines which context, and how much of it, should be removed for the component to have multiple uses.

DESIGN CONSTRAINTS THAT MAKE SYSTEMS TRUSTWORTHY

Most current software theory focuses on a program's static behavior by analyzing source listings because that is what can be analyzed. Little theory is posited on dynamic behavior and performance under load. Often we do not know what load to expect. Vinton Cerf, inventor of the Internet, has remarked that "applications have no idea what they will need in network resources when they are installed." As a result, we try to avoid serious software problems by overengineering and exhaustive testing.

Software engineers cannot ensure that a small change in software will produce only a small change in system performance. Industry practice is to test and retest every time any change is made in the hope of catching the unforeseen consequences of the tinkering. In one case, a three-line change to a 2 million-line program contained a single fault that caused multiple failures. The change was not fully tested as measured by a supplier's practices [15]. There is a lesson here. It is not faults but, rather, software failures that need to be measured to understand reliability.

Simplify the Design

A design simplification process eliminates elaborations that are a major risk to a project. "Gold plating," producing software embodying the most complicated interpretation of the requirements, occurs when designers have limited domain knowledge. They do not understand the very few places where generalizations are critical, so they generalize everywhere. It is also the consequence of not using prototypes during the requirements phase. Barry Boehm, another great software engineer, points out that "... only 60% of the features in a system are actually used in production" [16].

Software Fault Tolerance

Since failure is unavoidable in an imperfect world, the software design must be constrained so that the system can recover in an orderly way. Every software component and object class should provide special code that executes when triggered by a problem. A software fault-tolerant library with a watchdog daemon can be built into the system. When the watchdog detects a problem, it launches the recovery code peculiar to the application software. In transaction-processing systems, this often means dropping a transaction but not crashing the system. In administrative applications where keeping the database is key, the recovery system may automatically recover a transaction from a backup data file or log the event and rebuild the database from the last checkpoint and then continue transaction processing. Designers are constrained to define the recovery method explicitly for each process and object class using a standard library.

To implement the checkpoint technique successfully, one needs software functions of application monitoring, application failure recovery, application checkpoint/message logging, file replication, events logging and replay, IP packet dispatching, and IP address fail-over. A reusable library called Software-Implemented Fault Tolerance (SwiFT) [17] contains a set of reusable software modules for building reliable, fault-tolerant Windows NT, LINUX, UNIX, and Java applications. These modules can either stand alone or be integrated into existing software products to provide fault tolerance. Therefore, SwiFT is designed especially for object, process, and application replications using cold, warm, and hot replication schemes. SwiFT detects hang failures in addition to crash failures. These modules emerged from fundamental Bell Laboratories research in fault-tolerant software pioneered by Bernstein and Kintala [18].

Software Rejuvenation [19]

This constraint is to limit the state space in the execution domain. Today's software runs nonperiodically, which allows internal states to develop chaotically without bound. Software rejuvenation seeks to constrain the execution domain by making it periodic. An application is gracefully terminated and restarted immediately at a known, clean, internal state. Failure is anticipated and avoided. Nonstationary random processes are transformed into stationary processes. One way to describe this is rather than running a system for one month with all the mysteries that untried time expanses can harbor, to run it only one day 31 times. The software states are reinitialized each day, process by process. Decreasing the time between rejuvenations reduces the cost of downtime but increases overhead. One system deployed at more than 500 sites in the United States operated for two years with no outages, using a rejuvenation interval of one week. The following case history demonstrates the need for rejuvenation.

The Federal Aviation Administration's (FAA's) Voice Switching Communication System (VSCS) was upgraded in 2003 from UNIX to WinNT. WinNT brought with it the Microsoft problem of clock expiration after 49.7 days to prevent data overload, which then became an FAA problem that led to a massive failure. Harris Corporation, the WinNT VSCS software house, did not use rejuvenation technology that could have prevented this failure. Note that during VSCS development, the issue of WinNT as an industrial-strength, reliable platform raged in the telecommunications software trade press. Even though Microsoft had upgraded WinNT with clustering technology in a two-node failover configuration, industry skeptics argued against the risk of moving away from UNIX.

Harris's Web page reported as follows [20]:

> *The Voice Switching and Control System (VSCS) provide the Federal Aviation Administration (FAA) with a computer-controlled, highly distributed communications and control system to support air traffic management into the 21st century. The Harris-developed VSCS allows air traffic controllers to establish all Air-to-Ground and Ground-to-Ground calls for current and projected traffic volumes. The VSCS has completed design, development, testing, production and installation of the system for all 21 FAA ARTCC locations.... Software design and development, using SEI Level 3 methodologies; test plan development; software testing; data analysis; various legacy systems and tools.... VSCS is based on independent distributed processors and switches, fault-tolerant databases, redundant high-speed bus interconnections, and extensive switching for real-time reconfiguration and redundancy to achieve an operational availability of 0.9999999. Switchovers during fault detection, isolation and resolution are done without breaks in communications, so failures are transparent to the ATC.... [Harris delivered all the features to all 23 sites on on-time with exemplary quality]. Harris was the winner of the Contractor of the Year Award from the Human Factors Engineering Society for excellence in human-machine interface design. Customer quote: "In my association with Harris on the VSCS project, they have shown a dedication towards providing excellence. Their people are professional[s] who reflect pride in their company and their products."*

This was the correct strategy given the design and uncertainty of the traffic load. Shutdown is better than allowing an overloaded system to run and potentially give controllers incorrect information about flights. When we try to run software beyond its specified domain, we often fail in obscure ways. Greg Martin, the chief FAA spokesman in Washington, said that the failure was not an indication of the reliability of the radio communications system itself, which he described as "nearly perfect." Harris programmers were operating at or better than the state of the practice. They were SEI 3, and apparently the application was robust. The problem is that the software industry is neither aware of, nor using, available tools that would prevent many failures. Even worse, the same poor designs recur because software failures are rarely studied with the intention of teaching better methods.

Hire Good People and Keep Them

This might have been the first constraint because it is so important, but any software organization can adopt the three constraints mentioned above as they set about improving the quality of their staff. Hiring good people is not easy. Every shop claims to have the "very best' people"; obviously, very few actually could. It can take 16 weeks to bring a new hire onboard: 8 weeks to fill the job and another 8 weeks to train the person in the ways of the company and the specific project, all the while asking, observing, and checking to see if they care about trustworthiness.

The high correlation between defects in the software product and staff churn is chilling. Defects are highly correlated with personnel practices. Groups with 10 or more tasks and people with three or more independent activities tend to introduce more defects into the final product than those who are more narrowly focused. When Boeing improved their work environment and development process, they saw 83% fewer defects, gained a factor of 2.4 in productivity, improved customer satisfaction, and improved employee moral.

Limit the Language Features Used

Most communications software is developed in the C or C++ programming languages. Les Hatton's book *Safer C: Developing Software for High-Integrity and Safety-Critical Systems* describes the best way to use C and C++ in mission-critical applications. Hatton advocates constraining the use of the language features to achieve reliable software performance and then goes on to specify, instruction by instruction, how to do it. He writes: "The use of C in safety-related or high integrity systems is not recommended without severe and automatically enforceable constraints. However, if these are present using the formidable tool support (including the extensive C library), the best available evidence suggests that it is then possible to write software of *at least* as high intrinsic quality and consistency as with other commonly used languages."

C is an intermediate language, between the high and machine levels. There are dangers when the programmer can drop down to the machine architecture, but with reasonable constraints and limitations on the use of register instructions to those very few key cases dictated by the need to achieve performance goals, C can be used to good effect. The power of C can be harnessed to assure that source code is well structured. One important constraint is to use function prototypes or special object classes for interfaces.

Limit Module Size

The optimum module size for the fewest defects is between 100 and 1000 NCSLOC. Smaller modules lead to too many interfaces, and larger ones are too big for the designer to handle. Structural problems creep into large modules.

Initialize Memory

All memory should be explicitly initialized before it is used. Memory leak detection tools should be used to make sure that a software process does not grab all available memory for itself, leaving none for other processes. This creates gridlock as the system hangs in a wait state because it cannot process any new data.

Check the Design Stability

Software developers know that their systems can exhibit strange, unexpected behavior, including crashes or hangs, when small operational differences are introduced. These may be the result of new data, execution of code in new sequences or exhaustion of some computer resource, such as buffer space, memory, function overflow, or processor time. Fixes and upgrades create their own errors. The fact that the only recourse has been exhaustive retesting limits the growth of software productivity in enhancements to existing systems and modules. Experienced software managers know to ask "What changed?" when a system that has been performing reliably fails suddenly and catastrophically. Under current methods of software production, systems are conditionally stable only for a particular set of input and a particular configuration.

A software system is stable if a bounded input creates a bounded output. Instabilities arise in the following circumstances:

1. Computations cannot be completed before new data arrive.
2. Round-off errors build or buffer usage increases to eventually dominate system performance.
3. An algorithm embodied in the software is inherently flawed.
4. A soak-in period establishes the stability of a software system, and the failure rate decreases with time, even if no software changes are being made [21]. Failures after the stability period are caused mainly by configuration and administration errors, not by software bugs.

Bound the Execution Domain

Check that all inputs and outputs are reasonable. For example, the system should not accept a result of an altitude calculation that claims that an aircraft is at 67 feet off the ground when the plane is warming its engines on the ground. The domain experts specify a range of acceptable values for all inputs and outputs and the software needs to validate that the values fall within their defined ranges. When they do not, fault-tolerant software is invoked.

Let input data pass through an interface object to the processing software in a normalized and validated form. This isolates the processing programs from the idiosyncratic behavior of human operators and of the external environment. It reduces the need to change the application to stay in lock step with screen changes. If the resulting system cannot provide the required performance because of this indirection, it is best to buy a faster machine. This is a hardware problem, not a software problem. Should this argument fail, the architect must adopt *fast path* processing to bypass the overhead of indirection. The architecture becomes more tightly coupled and therefore more complex, causing the software to become more difficult to fix and extend. Fast path processing is used for just this purpose, to bypass the overhead of indirection, in many database access systems [22]. Bell Labs and IBM made this trade-off when they had serious performance problems because they could manage the inherited complexity.

Factor and Refactor

Factoring is the mathematical technique of finding common terms in an equation. Software designers need to look for common requirements, functions, and code throughout software development. Refactoring tweaks the factoring concept and applies it to software design. Refactoring defines the software technology aimed at reducing the size of the software by finding and eliminating redundant functions and code and dead-end code. Refactoring is the redesign of software in ways that do not change its functionality. Make it work, make it work right, and make it work better are the three phases of design. The idea is that the first and second iterations of software design and implementation stress understanding the feature, the problem domain, and getting the software to work. Refactoring is left to the third iteration—make it work better.

Refactoring can be used for small changes to improve structure incrementally, as suggested by Martin Fowler [23]: "Refactoring is a disciplined technique for restructuring an existing body of code, *altering its internal structure without changing its external behavior*. Its heart is a series of small behavior preserving transformations. Each transformation (called a 'refactoring') does little, but a sequence of transformations can produce a significant restructuring. Since each refactoring is small, it's less likely to go wrong. The system is also kept fully working after each small refactoring, reducing the chances that a system can get seriously broken during the restructuring."

Refactoring can also be used to indicate that a module must be rewritten, *keeping its functions and its interfaces constant while changing its internal structure*. One approach is "refactoring to patterns," which marries refactoring—the process of improving the design of existing code—with patterns, the classic solutions to recurring design problems. Refactoring to patterns suggests that using patterns to improve an existing design is better than using patterns early in a new design.

Allocating as much as 20% of the effort on a new release to improving the maintenance of the system pays large dividends by making the system perform better, avoiding failures induced by undesired interactions between modules, and reducing the time and space constraints on new feature designs. The goal is to reduce the amount of processor time old modules use and the amount of memory they occupy or input/output they trigger while holding their interfaces fixed. Other modules may be modified or new ones added to provide new features. This strategy naturally leads to reuse within the system. The greatest economic benefit is to reuse software at the application level.

Unfortunately, this process is not deployed widely because of the preoccupation with getting new features to market quickly. A bug-ridden store-and-forward system did use this concept through the 1970s and as a result grew to be extremely reliable. Rather than being tossed out and replaced with the next new thing, it continued to switch messages until 1995, when spare parts could no longer be obtained for the hardware. An unexpected benefit of "design for maintainability" is that new modules and those being upgraded for new features are more reliable because their developers do not face harsh space or time constraints. They could focus on getting their module to work while others concerned themselves with reducing the size of modules that did not contain new features.

CONCLUSIONS

Our industry is experiencing a sea change from being fast to being right. Trustworthiness is becoming a vital design issue. There are many proven ways to make software safer, more reliable, and more secure. When these architectural and design approaches are widely adopted, we project that there will be more robust systems and another division of labor into people who build components and people who integrate them into systems. Here is a checklist that summarizes many of the points of this chapter and is more fully explained earlier in this chapter and shown in the first reference:

Design for Reliability Checklist: Software

1. Simplify.
   a. Increase cohesion of modules and components.
   b. Encapsulate.
   c. Normalize interfaces.
   d. Reduce intermodule coupling.
   e. Reduce coupling in the build system.
2. Use fault-tolerant libraries and transfers for on-the-go recovery.
3. Rejuvenate the executing system from time to time.
4. Hire good people and keep them!
5. Limit the programming features.
   a. Use self-describing mnemonic parameter and variable names.
6. Limit the module-size initialize memory.
7. Check the design stability.
8. Bound the execution domain.
9. Engineer to performance budgets.
10. Reduce the algorithm complexity.
11. Factor and refactor.
    a. Allow one and only one function or operation per line of source code.
    b. Permit no negative, and especially no double negative, logic in the code.
    c. Keep specifications, design, source code, terminology, user interfaces, and documentation consistent.
12. Improve maintainability [with 20% of staff].
    a. Use uniform conventions, structures, naming conventions, and data descriptions.
    b. Optimize code for understandability while minimizing code complexity.
    c. Comment to explain extraordinary programming structures.
    d. Format code for clarity and understanding.
    e. Modularize architecture for configuration flexibility and growth.
    f. Store configuration parameters in a global database for ease of change.
    g. Harmonize all code after every change to assure compliance with the design rules.

REFERENCES AND NOTES

[1] Bernstein, L., and Yuhas C. M., *Trustworthy Systems Through Quantitative Software Engineering*, Wiley, Hoboken, NJ, 2005, Chap. 7. Details design issues with suggested solutions.

[2] Sha, L. Using simplicity to control complexity, *IEEE Software*, vol. 18, no. 4, July–Aug. 2001.

[3] Committee on Information Systems Trustworthiness, NIST, *Trust in Cyberspace*, National Research Council, Washington, DC, 1999.
[4] A fault is an external human error that becomes incorporated in the software. Failure is a state of no response to external stimuli due to the execution of a fault.
[5] http://www.nap.edu/readingroom/books/trust/.
[6] Parnas, D. L., On the criteria to be used in decomposing systems into modules, *Commun. ACM (Program. Tech. Dept.)*, Dec. 1972.
[7] Parnas, D. L., The influence of software structure on reliability, in *Proceedings of the International Conference on Reliable Software*, Los Angeles, ACM Press, New York, 1975; http//portal.acm.org/citation.cfm?id=808458.
[8] http://www.cs.caltech.edu/~adamw/papers/power-2.pdf.
[9] Man, F.-T., A brief history of TL1, *J. Network Syst. Manage*., vol. 7, no. 2, June 1, 1999, pp. 143–148.
[10] Meyers, http://hjem.get2net.dk/nimrod/tipdesign.htm.
[11] Liskov, B., Data abstraction and hierarchy, *SIGPLAN Not*., vol. 23, no. 5, May 1988.
[12] Sacchi, W., Collaboration practices and affordances in free/open source software development, in I. Mistrik et al., Eds., *Collaborative Software Engineering*, Springer-Verlag, Berlin, 2010, Chap. 15.
[13] The application integrators will use software manufacturing techniques explained in our book [1, pp. 392–398].
[14] Selby, R., Empirically analyzing software reuse in a production environment, in W. Tracz, Ed., *Software Reuse: Emerging Technology*, IEEE Computer Society Press, Washington, DC, 1988, pp. 176–189.
[15] http://ublib.buffalo.edu/libraries/projects/cases/computing/computing_ethics.pdf.
[16] Boehm, B., et al., *Prototyping vs. specifying: a multiproject experiment*, *IEEE Trans. Software Eng*., May 1984.
[17] U.S. patent 7,096,388.
[18] Bernstein, L., and Kintala, C., Components for software fault tolerance and rejuvenation, *AT&T Tech. J*., vol. 75, no. 2, Mar.–Apr. 1996, pp. 29–37.
[19] Trivedi, K., et al., On the analysis of software rejuvenation policies, in *Compass '97, Proceedings of the 12th Annual Conference on Computer Assurance*, June 16–19, 1997, pp. 88–96.
[20] http://www.harris.com (and search for VSCS).
[21] Jalote, P., Murphy, B., and Sharma, V. S., Post-release reliability growth in software products, *ACM Trans. Software Eng. Methodol*., vol. 17, no. 4, Aug. 2008.
[22] http://www-01.ibm.com/software/data/db2imstools/imstools/imsfastpathonl.html.
[23] http://www.refactoring.com; and Fowler, M., *Refactoring: Improving the Design of Existing Code*, Addison-Wesley, Reading, MA, 2000.

Chapter 14

Prognostics and Health Management Capabilities to Improve Reliability

Louis J. Gullo

INTRODUCTION

Prognostics and health management (PHM) capabilities are discussed in this chapter as they relate to reliable designs. Main topics covered include system health monitoring and health management for predicting failures and preventive maintenance actions, and improving design reliability with PHM through:

1. Preventive maintenance prior to failure
2. Condition-based maintenance to determine whether mission failure is imminent
3. Condition-based maintenance to predict failure occurrence
4. Corrective maintenance during and outside the mission window

Designing for prognostics capability ensures that failures are minimized during a mission when the product or system needs to operate failure-free. The prognostics capability accomplishes this goal by warning the user that a failure is imminent well in advance of the failure occurrence and manifestation of a severe or catastrophic failure effect. The prognostic capability is a

Design for Reliability, First Edition. Edited by Dev Raheja, Louis J. Gullo.

form of condition-based maintenance (CBM) and development of preventive maintenance action to correct the failure in a proactive manner instead of a reactive manner. Also, maintenance costs can be reduced by allowing maintenance to be performed during planned shutdowns or less critical times when a mission is not being executed. By using PHM to reduce maintenance costs, the system experiences the advantages of increased inherent availability and operational availability.

In fault-tolerant architecture and design, failures occur in the system and the mission can still be accomplished. Fault-tolerant design is usually possible through inclusion of redundant hardware and software elements in a serial–parallel system configuration. It is also possible to achieve fault tolerance through prognostics and health monitoring to determine when failures are imminent. This type of fault-tolerant design capability is normally accomplished by deferring failure repair and maintenance actions based on the robustness of the design to continue to operate in a degraded mode or with hot spare and redundant capability. Failures are imminent if they are going to occur in a given amount of time. Failures are deemed imminent due to the collection of health monitor data reporting stress events over time for statistical analysis to determine when failures are about to happen. During the collection of health monitor data, the mission can still be accomplished with a known probability of system failure and a risk probability that is deemed acceptable. However, if the system is allowed to operate too long, subsequent failures will cause the system to experience mission-stopping failures.

Interest has been growing in monitoring the ongoing health of products and systems to predict failure occurrences and provide warning of an impending catastrophic failure effect [1] and to reduce the need for corrective maintenance. The deterioration of product and system health is defined as the degradation or deviation of performance from normal operating conditions. Prognostics are the prediction of the future state of health based on current and historical health conditions [2].

The PHM system is successful if the system health monitoring capability predicts when failures are going to occur before the failures actually manifest or if the failure mechanism surfaces without resulting in a mission-critical failure. The failures that are imminent and have not occurred may be failures due to stress conditions over time. As time passes or the stress increases, the fatigue on the component increases, as does the likelihood of failure. The accumulation of fatigue continues until the operating point, which results in physical damage that requires a fix or a replacement to remedy the problem. This physical damage may be caused by thermal, environmental, mechanical, electrical, chemical, or metallurgical conditions.

Even though recently published literature [1] suggests that "prognostics and health management capability is well established for various mechanical productsor system applications, but is a relatively new concept for electronic products and systems," this is not the case. Both mechanical and electrical products, as well as several system applications, have used PHM design features which appeared many years ago. PHM electrical design features for large military and aerospace systems are not new technology, nor are they recently deployed. PHM design features have been receiving increased visibility and renewed interest over the last several years, due to added emphasis on remote maintenance monitoring and testability design characteristics for long-distance support. There are many examples of electrical products and system applications with PHM design features. Two examples are given where PHM was implemented in electrical designs, one from the distant past and one current.

The first example explains how PHM was implemented in a program that existed in the 1970s. We use the name "program XYZ" to refer to this example. For program XYZ, it was imperative to design it right the first time. Program XYZ could not afford to spend money on costly prototypes that were designed improperly and create excessive engineering scrap. Program XYZ wanted to model and simulate everything prior to building the first engineering models, and to transfer the design seamlessly from advanced development to production without scrap and waste. Two of the critical functions of the system for program XYZ were the measurement of testability parameters related to the probabilities of fault detection and isolation, described later in the chapter. The assessment and verification of these parameters were part of the program that determined the effectiveness of the built-in test (BIT) logic. Special tools were created to exercise the logic using various models and simulations. Even though PHM was not what the capability of the design was called back then, PHM was fully implemented on program XYZ beyond what is typically seen in today's system of systems integration (SOSI) designs.

The second example is related to a current PHM technology on an existing large-scale military SOSI design and development program. We refer to this example as "program ABC." Program ABC is a SOSI design on a larger platform than that of program XYZ, but not by very much. Program ABC uses BIT signals similar to those used on program XYZ, except that now the technology is much faster, more mature, and more robust. The integrated platform uses PHM features and capabilities in places where predecessor designs could have used PHM, but chose not to, due to cost constraints. PHM is a technology differentiator on program ABC's SOSI platform compared to predecessor SOSI platforms of an equivalent class and nature. With this SOS platform, program ABC has the potential to showcase the latest in PHM technology, along with the latest condition-based maintenance (CBM) capabilities as long as the funding stream continues as planned to build the number of platforms originally intended.

PHM IS DEPARTMENT OF DEFENSE POLICY

Although PHM implementation in electronic products and systems is not new, there is increased emphasis today on its value and importance in several marketplaces, such as the military market. For example, the Department of Defense (DOD) recently established a policy called condition-based maintenance plus (CBM+). In DoD Policy 5000.2 [3], within performance-based logistics (PBL), it states: "... optimize operational readiness through affordable, integrated, embedded diagnostics and prognostics. ..." The CBM+ policy was derived to improve design reliability, inherent availability, and operational availability, and reduce logistics life-cycle costs that improve the total cost of ownership. CBM+ minimizes unscheduled corrective maintenance by warning and predicting the imminent failure. With the prediction of the failure, a potential latent failure mechanism, preventive maintenance is scheduled at an opportune time in the near future prior to the failure mechanism propagating into a patent failure. By replacing an item prior to the detection of its patent failure, design reliability, inherent availability, and operational availability are improved. This improvement occurs since the item is replaced at a time when the system or product can be taken off-line to perform the preventive maintenance. The preventive maintenance is performed at a time prior to the occurrence of a mission-critical failure. When a mission-critical failure occurs, corrective maintenance must be performed to restore online system functionality and mission-essential operations. If the risk of a corrective maintenance action can be deferred to a scheduled preventive maintenance servicing, then reliability and availability are not affected by the mission-critical failure. Only corrective maintenance actions affect reliability and availability in a negative way.

CONDITION-BASED MAINTENANCE VS. TIME-BASED MAINTENANCE

CBM relates to maintenance actions driven by the condition of the equipment prior to physical indications of failures that must be repaired immediately to place the system back into service or to restore the product back into operation. Time-based maintenance (TBM) is scheduled maintenance planned way in advance of launching the product to market or prior to fielding the system and placing it in service. Both CBM and TBM are forms of preventive maintenance. In the case of CBM, the maintenance is not planned or scheduled. CBM occurs only when the conditions of the equipment warrant the necessity to perform hardware replacement and determine the best time to conduct maintenance actions. It is important to note that preventive maintenance actions such as CBM or TBM do not affect design reliability. When CBM is used in lieu of corrective maintenance, there is an immediate improvement in inherent design reliability and availability since the implementation of PHM, if done correctly, will ensure that failures will not occur and that corrective maintenance will not be needed during the time when the product or system is required to operate.

PHM as Part of Condition-Based Maintenance

Next, we discuss how PHM works with an added emphasis on testability. PHM combines the design applications of embedded sensors, remote sensors, BIT signals, fault detection, fault isolation, performance trending, analog and digital signal processing, status and warning messages to a display, and fault logging as part of a larger data reporting schema. Performance trending with analog and digital signal processing could include interpretation of thermal, vibration, stress, strain, functional, and performance parameters that are measured and stored as a series of cumulative distribution functions over time. These parameters could be any design characterization that warrants monitoring to assess product or system health. Just as the human body has a central nervous system to trigger discomfort or pain when physical damage or an illness occurs, warning the person to seek medical attention, the same should be true for electronic systems with PHM capability, where PHM capability can be afforded. Based on a product's or system's health, determined by its monitored conditions, maintenance procedures can be developed and scheduled in advance of failure detection.

Two different approaches to PHM are (1) monitoring and reasoning of failure precursors, and (2) monitoring environmental and usage loads for damage modeling.

MONITORING AND REASONING OF FAILURE PRECURSORS

A failure precursor is a device that signals an imminent failure event. A precursor indication may be a signal which normally monitors a design parameter that suddenly detects a change in the measured value so that a failure trend is determined. For example, a sudden increase or decrease in the output voltage or output current of a voltage regulator, even though the parameters are still within the manufacturer's ratings for the regulator, probably indicates the presence of a failure mechanism that if uncorrected would result in a product or system failure. Expendable electronic or mechanical devices such as fuses and canaries have been used for electronic system PHM for many years. Fuses provide circuit protection from high electrical stress conditions that may occur infrequently. Canaries provide advanced warning of imminent failures, which may be caused by wearout failure mechanisms.

MONITORING ENVIRONMENTAL AND USAGE LOADS FOR DAMAGE MODELING

The customer use applications for a product or system will consist of operation profiles or mission profiles over the product life cycle or system life cycle where loads are varying and transient conditions occur that place unwarranted stress on a product or system. These stress occurrences may be construed by the product or system designer as externally induced electrical or mechanical overstresses that are beyond the limits of the product or system, possibly outside the specification requirements or the operational capabilities of the design. These stress conditions, which may occur either individually or in a combination of stress conditions simultaneously, may initially degrade performance or cause accumulated fatigue through gradual physical degradation until a failure mechanism develops into a patent failure.

There are a minimum of three states that could be used to describe PHM for a system. For example, the three states could be: (1) health is good; (2) health is bad; and (3) health is questionable. The three states could also be defined by the use of red–yellow–green states as outputs from condition monitoring circuitry. A red state symbolizes that a patent failure has occurred and requires corrective maintenance, a yellow state symbolizes that an imminent failure has been detected with a corresponding probability of occurrence and time to failure and requires preventive maintenance in the near future, and a green state is fully operational. The yellow state includes degraded operation and degradation modes where operating parameters drift away from their expected values, failures of redundant elements in a fault-tolerant architecture where the system mission is still accomplished, and failures of backup systems, with primary systems still running.

FAULT DETECTION, FAULT ISOLATION, AND PROGNOSTICS

The capabilities of a product or system design to detect failures within itself automatically are usually called its built-in test (BIT) capabilities. As a minimum, all BIT designs should be able to monitor the health of a product or system and diagnose undesirable failure events. BIT design will be implemented onboard as embedded hardware and software diagnostics. BIT signals that continuously report the health of a system and diagnose the system should also be able to identify, locate, and report the occurrence of a failure. Once a failure is detected, the system should also have the capability to isolate the failure to a replaceable assembly in order to initiate a repair action, for corrective maintenance. The BIT circuitry may also consist of error detection and correction capability. This means that the design will be programmed with logic that finds, analyzes, and fixes errors and problems before they become higher-level failures, without intervention from a person such as a maintainer or operator. This error detection and correction feature may include self-checking circuits and software, or self-verification circuits and software. The assessment of the capabilities of a system to perform fault detection, fault isolation, and prognostics may be classified as maintainability engineering or testability engineering. These two specialty engineering disciplines consider the design features necessary to ensure that fault detection, fault isolation, and prognostics requirements are met. They assess the capabilities of products and systems to ensure that requirements are verified.

Typical requirements might be probability of fault detection at 95% for all failure modes of all line replaceable units (LRUs) or field-repairable units (FRUs), and probability of fault isolation at 95% for an ambiguity group of three for all failure modes of all or FRUs. An ambiguity group of three means that three LRUs are identified as potential causes of the failure. For prognostics, a requirement may be written which requires all LRUs or FRUs to report any high-stress conditions 100 hours prior to detection of 95% of any failure modes. The same BIT design capability used to detect and isolate faults may be able to detect and isolate imminent faults for prognostics capability by continuously monitoring the condition of the critical signals in situ to include design parametric stress measurements and data from sensors. BIT signals, whether or not they are designed into a system for PHM, may be classified into three types:

1. A-BIT or S-BIT: automatic BIT or startup BIT, initiated on startup of the product or system, and usually involves a fully functional test to ensure that all critical requirements are met. An A-BIT is not as thorough in terms of test coverage as an I-BIT, but more thorough than a C-BIT. There is usually no difference in A-BIT designs for a system with PHM or without PHM capabilities.

2. C-BIT: continuous BIT, which runs in the background during normal product or system operation, and usually is a cursory set of tests requiring minimum processing power to ensure that the majority of critical functions are operating. The C-BIT is the least amount of test coverage of the three types of BIT. The C-BIT for a PHM system will have more design analysis features than will a system without PHM. The design analysis features for a C-BIT in a PHM system will include statistical data analysis and prognostic reasoners to detect when a failure is imminent.

3. I-BIT: initiated BIT, started manually by a maintainer or operator during normal system functions by exercising specially programmed diagnostics and prognostics to determine causes of errors, faults, or failures, and to assist maintainers in the fault detection and isolation of failures. An I-BIT is the most thorough form of BIT, with the highest percentage of test coverage or fault coverage. When running an I-BIT, normal system operation may be suspended or may be slowed down to allocate processing power to the enhanced BIT programs. In an I-BIT, maintainers should have access to C-BIT and A-BIT logs where errors and fault data are stored and easy to retrieve for further analysis of the product or system. There is usually no difference in an I-BIT for a system with PHM or without PHM.

SENSORS FOR AUTOMATIC STRESS MONITORING

Utilization of A-, C-, and I-BITs may be accomplished with the implementation of two types of sensors and monitors: mechanical and electrical.

1. *Mechanical sensors*. Thermal sensors (such as thermocouples), vibration sensors (such as accelerometers), stress gauges, and strain gauges are a few examples of mechanical sensors.
2. *Electrical sensors*. Voltage sensors, current sensors, charge sensors, magnetic sensors, impedance sensors (including resistance sensors to determine opens circuits, short circuits, and low or high resistance, and reactance sensors to determine low or high capacitive or inductive parameters), power and power density sensors, frequency sensors, noise sensors, and timing sensors are a few examples of electrical sensors.

The cost justification to add these mechanical and electrical sensors into a product or system design must be justified by the return on investment. The cost justification varies according to the various product markets and to the need for highly reliable products and systems that are proven to perform in high-stress environments and conditions. Medical, computing, commercial power grid, power distribution, commercial flight avionics and aerospace, commercial automotive products, and military systems are all examples of high-stress, high-reliability systems with the need for design requirements that specify PHM technologies.

In the case of DoD CBM+ policy, there is an increased awareness throughout the military acquisition community that PHM is important and saves the government money, but it does not come without a cost. In performing cost trade studies between PHM design implementation costs and life-cycle cost avoidance through increased inherent and operational availability, the cost for design implementation may be too prohibitive when considering the overall total design to unit production costs (DTUPC). Even though the DoD goal is to implement CBM+ and PHM to the maximum affordable extent possible, the DTUPC budgets must be considered.

Whether a design program is DoD or nonmilitary, specific maintainability, reliability, and test coverage requirements are established. These requirements must be considered in comparison with other system engineering and product engineering requirements while meeting all constraints. Assessment and trade-off studies of design alternatives to resolve multidimensional problems consider each alternative and select the option that maximizes the implementation of the requirements, balanced by the available funding and given target costs without compromising the key performance factors, which may include reliability, operational availability, operator and maintainer manning, volume/dimensions, and weight. In other words, get the biggest bang for the buck in spending money for PHM capabilities.

REFERENCES

[1] Chapter 67 titled: Prognostics and Health Monitoring of Electronics by Vichare, N., Tuchband, B., and Pecht, M., *Handbook on Performability Engineering, edited by Krishna B. Misra,*, p. 1107, published by Springer, 2008.

[2] Vichare, N., Tuchband, B., and Pecht, M., Prognostics and health management of electronics, *IEEE Trans. Components Packag. Technol.*, vol. 29, no. 1, 2006, pp. 222–229.

[3] *Defense Acquisition Guidebook*, DoD Policy 5000.2, U.S. Department of Defense, Washington, DC, Dec. 2004.

Chapter 15

Reliability Management

Joseph A. Childs

INTRODUCTION

In this chapter we provide both motivation and guidance in outlining the importance of reliability and the techniques to assure optimal design program success. If the reliability of a product is considered one of its key performance characteristics, activities to assess and improve the system or product reliability must receive the time and resources necessary to assure success. These activities help to minimize the risk of program delays and overruns due to design issues, test anomalies and mishaps, and field failures. Reliability activities early in the design process and employed continuously throughout the product life cycle must be considered crucial to minimize or eliminate the chance of program failure.

A Case for the Importance of Reliability to Management

The following are excerpted comments from the Nobel laureate physicist Richard Feynman from his notes regarding the *Challenger* shuttle catastrophe [1]:

> From Introduction: It appears that there are enormous differences of opinion as to the probability of failure with loss of vehicle and human life. The estimates range from roughly 1 in 100 to 1 in 100,000.... Since 1 part in 100,000 would imply that one could put a Shuttle up each day for 300 years expecting to lose

Design for Reliability, First Edition. Edited by Dev Raheja, Louis J. Gullo.

only one, we could properly ask, "What is the cause of management's fantastic faith in the machinery?"

From Conclusion: ... If a reasonable launch schedule is to be maintained, engineering often cannot be done fast enough to keep up with the expectations of originally conservative certification criteria designed to guarantee a very safe vehicle.... They therefore fly in a relatively unsafe condition, with a chance of failure on the order of a percent.... Official management, on the other hand, claims to believe the probability of failure is a thousand times less ... demonstrating an almost incredible lack of communication between themselves and their working engineers.

And finally: For a successful technology, reality must take precedence over public relations, for nature cannot be fooled.

Feynman cites many instances where the engineering-to-management chain of communication might have been inadequate. The point here is not that National Aeronautics and Space Administration management is inadequate or that its engineering organization is less than excellent. Rather, it is important to note that it is *necessary* for engineering personnel to communicate with management in a manner that is honest, useful, and as complete as possible. It is also *imperative* that management use the insights it gains from engineering staff to face realistic challenges and respond with the proper funds, scheduling, and scope changes that help assure success. In almost any design project, there will be a tension between performing a complete job with safety and reliability at the top of the performance list against the triple pressures that management must face every day: performance, schedule, and budget. "Nature cannot be fooled."

Management of a Reliability Program

The most important goal of any reliability program is to affect the design and minimize the risks of critical failures of the design's life cycle. Improving the design is why the assessment, analysis, and test tasks are performed. Completion of the tasks provides no benefit to program management unless the results are used to improve the design, and to better understand and address reliability risk if design improvements cannot be implemented.

For management to be able to use these results, the risks to design reliability need to be presented in terms of costs, schedule, performance, the customers' needs, and marketing the competitive advantage for the product. Improving the design and understanding the risks cannot occur without the reliability program being a robust and integral part of the overall program.

Since the main goal of a reliability program is to assess and affect the design, this portion of a design program is actually a subset of the design effort itself. Certainly, the program manager is responsible for meeting the reliability goals, just as the other key program parameters, such as schedule, budget, and other performance requirements, are the manager's responsibility. However, the onus for achieving reliability goals is on the design team, including the designers and design support entities. The interfaces between the involved organization is extremely important to the success of the reliability program. These entities include mechanical (thermal and dynamics) engineering, electrical engineering, software engineering, quality assurance, procurement, and other organizations. The likelihood of success is dependent on the design manager's leadership in this effort, along with that of program management. If the design manager has the attitude that the reliability of the design is as important as the rest of the key factors, those goals will almost certainly be manifested in project decisions throughout the program life.

The following is a description of the tasks involved in managing reliability and a reliability program in the design and development of a new product. It refers to many subjects covered in much more detail elsewhere in this book, but with a difference. The emphasis here is on how these tasks dovetail with other efforts being performed, and how the interfaces can work to produce a synergism between the reliability professionals and the rest of the program. Figure 1 is a representation of such dovetailing. For example, during the concept design of a product, the design team would perform initial reliability predictions, parts selection, stress analysis following derating guidelines or rules, and design failure modes and effects analysis (D-FMECA). D-FMECA complements the design effort because findings early in the program could save much time, effort, and money, avoiding costly redesigns or rebuilds in the later phases. More details are provided in Chapter 4, which covers how design tools can be used and when such efforts are optimal in each product design phase.

The reliability program is like any other program: with planning, implementation, controls, and reporting. As would be expected, they are all performed in concert with the design and development programs to develop a seamless, coordinated effort. IEEE Reliability Society Standards [2] are excellent references with insight into performing the reliability tasks, from both a technical and a managerial viewpoint.

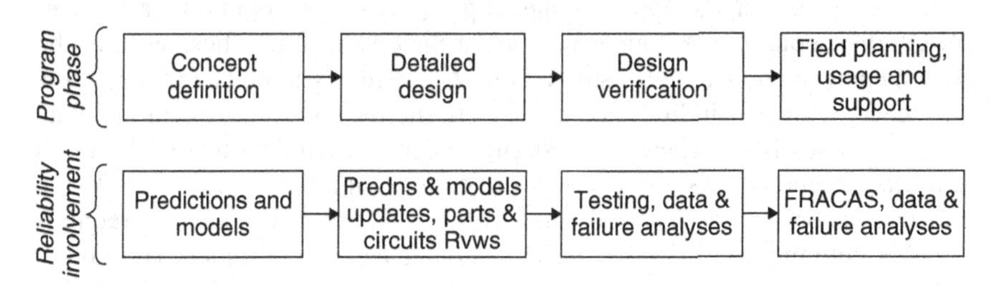

Figure 1 Reliability engineering involvement in design, start to finish.

Reliability: A Key Element in Product Life (See Figure 1)

Important Early Program Contributions

At the beginning of a program, or during the proposal stage, there are key questions that must be answered. Not addressing these issues invites many problems and dissatisfaction later, when time and resources are at a premium. Examples of such concerns are:

1. What are the reliability requirements: while in use or in mission; while being transported or stored; or while being tested?

2. Can the reliability requirements be achieved and verified?

3. Is a design margin needed to address variation in customer operating conditions, overstressing the limits of the design, and factors leading to the potential for product abuse and mishandling?

4. What are the warranty and end-of-life requirements?

5. Will there be special materials that need specific planning for disposal? An example of such an issue is the treatment of nuclear waste materials when a nuclear power plant is shut down. Another case is a new requirement being developed in Japan and China for products that contain lead, such as solder in electronic packages. The standards being considered in those countries require the manufacturer to provide for disposal of such materials to mitigate environmental impact.

6. What is the definition of failure? This may seem obvious, but on many occasions the design organization believes that the product meets requirements, but the customer is dissatisfied. The design team is surprised to discover that the product is being used by a customer in a way that was not intended. Another example of such "surprises" is the system behaving in a manner different from that expected although the customer is using the product as directed. The product may perform in a probabilistic manner rather than a deterministic manner. This type of intermittent product behavior or erroneous event represents a "no win" situation for the design team. On one hand, the design team could be "right," avoid the customer's perspective and opinions, and completely disappoint the customer (perhaps at the cost of future business); or the design team can bow to the customer's wishes while assuming additional costs and schedule responsibilities and risks. Defining the term *failure* will facilitate a win–win situation, where the customer receives a product that performs as expected, keeping costs and resources in control, and the designer can design and test according to clearly defined criteria. A win–win situation requires detailed communication between the customer and the designer. The failure definition must be objective, definable, and reasonable. Both parties should consider details of what a failure is and what it is not, and use examples to define such failure events. Some examples of poorly defined failure definitions from the customer's perspective:

a. Behaves in an unexpected manner
b. Knocking sound
c. Coating finishes changing the texture
d. Fails after being used in an unusual or abusive way
e. Problems after storage under harsh conditions

These examples could all be considered issues according to a customer but not be adequate for the supplier's design team. They become well-defined failures if they are defined in a written agreement, specification, statement of work or standard, or other guiding document. The poorly defined failure examples provided above should be *treated* as failures, even if they aren't clearly defined. The supplier must respond to customer dissatisfaction and anticipate how the product will be used to avoid such issues, if possible.

7. What is the cost of the reliability program? Often, key subtleties in estimating the cost of a reliability effort are not addressed directly. For example, when reliability analyses and tests are performed, the systems test and detailed design engineering organizations are involved directly in responses to identified events and findings. One way to estimate budget and track costs is use of a *work breakdown structure* to categorize the responsibilities, skill levels, travel, material, and other costs involved in each entity.

8. What are the cost benefits of a reliability program? Another key element in considering reliability costs are the benefits. One key benefit that can be given short shrift is avoiding or minimizing the cost risks of failures. This is a particularly difficult issue to address, because it is very difficult to put a number on the quantity and severity of failures that may be avoided or reduced—the costs of not only of repair but also of customer dissatisfaction, delays, and redesigns that might be required. Accounting for warranty costs, avoidance of damage and safety risks, repeat business concerns, and other life-cycle costs defines substantial savings that can be realized. These are not typically considered in estimating the cost of a reliability program, because they are unknown risks. However, this should not prevent consideration of these issues, using experience, analyses, and judgment to quantify these gains to provide a balanced view of the costs and benefits.

9. How is the product to be modeled? What are the highest risk elements in the program? How can they be addressed early? These questions require reliability analyses in the concept and early design phases of a project: predictions, mathematical modeling, and insight into the actual design to be performed. The predictions and models are very dependent on the definition of failure, the performance, and the reliability requirements. Often, there is no detail to the product being designed. In these cases, knowledge of the design plans can help the product to undergo a first-order analysis by use of similarities to other designs while factoring in risks associated with new technologies or approaches. A subtle but important task is to review new parts, materials, and processes planned for use, and define possible actions to evaluate and address potential problems. These issues tend to hit new design programs the hardest early in their development.

Key Interfaces During Design

- *System design*. Assure that specifications are reasonable, verifiable, and complete. This includes not only the product or design specifications, but also those provided for the multiple tiers of suppliers, as appropriate. A clear understanding of the system and its performance requirements is critical to effective execution of the design project.
- *Detail design*. This includes electrical, mechanical, software, and systems engineering. Is system redundancy required? Special parts, derating, cooling, vibration effects, problem solving (including root cause analysis, improvement implementation, risk assessment and avoidance) should be considered.
- *Quality and production engineering*. Production risks should be understood, especially related to new part types, processes, and materials.
- *Procurement assurance (purchasing)*. Include reliability-related requirements for suppliers to perform FMECAs, reliability assessments, and problem solving regarding potential issues. Margin must be included in the requirements so that there is failure-free performance over the expected life cycle. This includes a clear definition of the operational environment and stresses. The suppliers need to perform risk assessments, including the identification of critical design features that must be closely controlled.
- *Test engineering*. The reliability team must be part of the definition of reliability-related test requirements (ESS, qualification testing and Analysis, and special design environmental tests). Examples of the special environmental tests are: design validation testing, HALT (highly accelerated life testing), and other accelerated life testing.
- *Management*. The program and technical management characterizes and prioritizes the risks. Reliability program planning and completing and reporting tasks and risks to assure close communication among the interfaces listed above are part of the reliability management effort.

Key elements of communicating with interfaces are listed below. In some circumstances, such elements are included in programs as part of doing business. For example, in military and space programs, formal reviews and analyses are typically required. Even in industrial or commercial programs, as a prudent business practice, these elements (e.g., toll gates) should be incorporated as part of a sound product development process.

- *Design reviews and analyses*. The point of design reviews and analyses is to consider design risks across multiple functions, including between hardware and software. This allows a synergy of thought that develops in a more complete and focused list of possible risks, as well as actions to address them. Subjects addressed in these reviews and analyses include parts stresses, system-level hardware and software trade studies and compromises, program priorities, and new or untested parts, materials, or processes. D-FMECA results make excellent source material for such reviews.
- *Co-located team*. Many issues, trade-offs, and concerns are discussed "in the hall" among designers, packaging personnel, managers, purchasing staff, and others; that is, they are covered informally between engineers and other entities. Sometimes this is where important initial decisions or events are discussed, so it is very important that the reliability function be co-located with such co-workers to facilitate being part of the design team. Conversely, the reliability function must be proactive in contributing to the design team.
- *Management involvement*. When a reliability issue arises, the program and technical leadership must consider this as a program performance concern, since the reliability of a product *is* embedded in the long-term performance of that product. Such reliability issues become a trade-off among the performance requirements, customer satisfaction, budget, and schedule. Skipping or reducing the scope of key reliability tasks, such as reliability testing, can result in increased risks, costs, and delays later at more critical times in the project or program.

PLANNING, EXECUTION, AND DOCUMENTATION

Reliability Program Plan

As the plan for an overall program is developed, it is important that a plan for the reliability program within that effort be defined. This requires that the scope, the resources, the budget, and the schedule be dovetailed into the overall program. Because reliability focuses on the risk of failures and unforeseen technical issues, this portion of the program should be included in design contingency planning and budgeting as well. These considerations help to assure that the reliability function can operate effectively in the program and provide the needed information, factors, and recommendations for a complete picture of the product as it is created, developed, and matured over its lifetime.

Reliability Program Tasks

The person or team responsible for implementing the reliability program must participate in tasks that directly involve or affect the reliability of the product and how reliability will be addressed:

1. Specifications reviews. This is both for the product and for components supplied to make the product.

2. Design analyses and reviews. The design must be analyzed and reviewed not only for its performance and quality requirements, but for steps that have been taken to assure that it meets the program or customer needs for lasting over time in operation, storage, shipping, handling, and maintenance.

3. Supplier issues. The supplier must participate in discovering and addressing key concerns that affect reliability for the item supplied. An effective failure reporting and corrective actions system (FRACAS) is a powerful tool in anticipating and addressing such concerns, especially for failure trends. Such a system can help shine light on what the actual underlying cause of a failure or set of failures is.

4. Production issues. The reliability team must address concerns related to packaging, assembly, and manufacture. If a product is difficult to build or requires new processes to manufacture, experience shows that such features commonly turn into reliability as well as quality issues. If a product is difficult to build, this can often lead to unexpectedly poor reliability or quality. Such issues can be addressed by developing a more robust manufacturing process, but more often than not the simplest action can be to alter the design to improve the product's producibility. This effort is focused on removing the possibility of assembly errors, such as backward diodes, improper or no torquing, or too much or too little heat in a solder profile. Tooling, clearly defined tolerances or actions, pictures in instructions, special fixturing, and other "fool-proofing" methods can help, but the point is that design and manufacturing personnel communicate with each other before a new part, process, or material (PPM) is selected. In some industries, where reliability is paramount, PPM guidelines are an intrinsic part of the design rules from the start of a program. Process FMECA has been developed specifically to help anticipate such concerns.

5. Reliability and other environmental tests are not only important but are often necessary to performance and delivery success. That is one reason why it must not only be performed, but performed early enough in the project to affect the design *before* it is committed to be produced and provided to customers. One trap is to perform reliability testing too late in the program to address economically the issues found. This diminishes the advantages of reliability growth while increasing the cost of incorporating improvements. If the testing is done after deliveries to customers, the program cannot avoid the "hidden costs" of customer dissatisfaction and often, increased cost and schedule risks.

6. Root-cause analysis and corrective actions. When failures, failure trends, or manufacturing problems are found, it is imperative to learn how the failure occurred. Fault trees are particularly effective in developing possible and actual causes of an issue. They help to facilitate the determination underlying actionable causes that can be corrected. Sometimes the cause is due to lack of robustness in the design, but often it is related to how the product is manufactured or assembled, as mentioned above. If the issue is related to the part being supplied, the supplier must be added to the team to help solve and address the causes. Root-cause determination is extremely important, because its results lead to optimal choices in improving the robustness of the product.

Reliability Documentation and Reporting

1. Reliability assessment (stress and environmental analyses, predictions, reliability modeling, and parts selection). These tasks can affect the design directly by anticipating possible issues when the design is not yet "cast in concrete" and alternative approaches can be considered without prohibitive costs or schedule slippage.

2. Assessment updates (update of reliability analyses and predictions, risk and problem assessment). More often than not, a design program is in a state of flux. That is, the design changes as more is known about the requirements, the manufacturing and testing, the parts and materials available, and for that matter, how the product will be used. When the design is updated, the reliability performance may change as well. Updating the assessment of those parameters is important to assure that no "catastrophic improvements" have been implemented! If issues do arise, it is better to know earlier than later.

3. Test reporting, including findings and reliability improvements. It is always good management practice to measure important characteristics—not only costs and schedule, but performance, including reliability performance. Reliability performance can be measured by reliability growth plots, trend analyses, and listing the "lessons learned" as a result of testing. Reporting these factors in a formal report is one way to assure that those lessons are captured for future design and test programs. Such lessons should be considered for incorporation into design standards and training, as warranted.

4. Failure reporting and follow-up reports. Included in all testing activities is the specter of failures that may appear, whether they appear during "breadboard" functional checks, formal software/hardware performance evaluations, verification and qualification tests, acceptance/screening test, or field experience. An excellent tool in any reliability program is a failure reporting and corrective action system (FRACAS). This process includes a taxonomy that captures what and when equipment fails, what the causes of such failures were, and how they were corrected in the long run (i.e., not *repaired* but *corrected*). With a well-designed FRACAS methodology, failure data can be sorted and viewed to provide engineers and managers with useful pictures of trends and failure drivers. This helps not only in performing root-cause analysis for existing failures, but can lead to important lessons, as mentioned above.

CLOSING THE FEEDBACK LOOP: RELIABILITY ASSESSMENT, PROBLEM SOLVING, AND GROWTH

The Feedback Loop

The reliability program provides a feedback loop to the product design effort. When assessing the reliability, the program can predict possible areas of high failure rate, provide warnings of overstress or overheating that should be addressed, and help select parts that are most likely to survive the environments required for the product. Reliability testing provides insight into the design reliability and its margins. Review of failure data and analyses throughout the program can provide knowledge of the underlying causes of failures. All this information is fed back to the program design and management functions. As issues are addressed, further analysis and testing provide more feedback regarding the effectiveness of the actions. Ultimately, the product's ability to withstand stresses improves. It provides true benefit to the program by helping to avoid surprises, resulting in cost and schedule overruns and the need for costly redesigns or product changes later in the product life cycle. In a larger sense, this same feedback should be applied to the design process itself to prevent systemic issues that can appear in multiple programs. In other words, why would a particular design issue occur in the first place, and what can be done to prevent reoccurrence of the same design issue in other programs or products?

The Language of Managers

Any decision in a program—even a technical one—is a business decision. If a program manager is to decide how to assess, avoid, or address a reliability issue, that manager must have business facts as well as technical facts. The quality guru Joseph M. Juran sometimes referred to this as the "language of money" [3]. That is, it is the responsibility of the engineering community, including the reliability engineer, to address reliability issues in terms of the costs to implement and verify, as well as the costs that would be avoided. It includes addressing resources and schedule concerns. These costs are not just in terms of money. Certainly, schedule and other drivers must be considered as well. This allows management to trade-off the various factors to make an informed decision.

However, for that decision to be sound, the manager must account not only for the "cost of failure" but also for the benefits of "failure avoidance," including hidden costs or risks, such as loss of customer confidence (image), legal actions, schedule slippage, and competitive advantages with better reliability. A high priority should be assigned to reliability assessment and improvement tasks to assure that the design reliability grows to meet or exceed the reliability requirements, as well as all the other program requirements. Another part of the cost trade-off equation is the benefit of reduced maintenance and logistics costs. This is a *big* selling point in earning customer loyalty!

Management and Reliability

In this chapter we provided both motivation and guidance in outlining the importance of reliability and the techniques to assure optimal design project success. As long as the reliability of a product is considered one of its key performance characteristics, it will almost certainly receive the time and resources necessary to assure design project success.

REFERENCES

[1] Feynman, R. P., personal observations on reliability of shuttle, in *Report of the Presidential Commission on the Space Shuttle Challenger Accident*, Vol. 2, App. F, U.S. Government Printing Office, Washington, DC, July 14, 1986.

[2] Several IEEE standards; for example:

[3] *IEEE Standard for Organizational Reliability Capability*, IEEE 1624–2008, IEEE Reliability Society, Piscataway, NJ, Feb. 2006.

[4] *Reliability Program for the Development and Production of Electronic Products*, IEEE 1332–1998, IEEE Reliability Society, Piscataway, NJ, June 1998 (reaffirmed 2004).

[5] *IEEE Standard Framework for the Reliability Prediction of Hardware*, IEEE 1413–2010, IEEE Reliability Society, Piscataway, NJ, Apr. 2010.

[6] *IEEE Guide for Selecting and Using Reliability Predictions Based on IEEE*, IEEE 1413.1–2002, IEEE Reliability Society, Piscataway, NJ, Sept. 2002.

[7] *IEEE Recommended Practice on Software Reliability*, IEEE 1633-2008, IEEE Reliability Society, Piscataway, NJ, June 2008.

[8] Juran, J. M., *Juran on Quality by Design: The New Steps for Planning Quality into Goods and Services*, Free Press, New York, 1992.

Chapter 16

Risk Management, Exception Handling, and Change Management

Jack Dixon

INTRODUCTION TO RISK

Risk has been defined, as the possibility of loss or injury. Risk is present in all human activities; there is risk in everything we do. There are many different types of risk that concern risk managers. A common thread among these risks should be the involvement of risk managers to assess the risks, to determine the future effects if the risks materialize, to develop contingency plans, and to select the best choice among alternatives to mitigate or eliminate the effects of the risks assessed.

Since risk measures a potential loss, it can be viewed as an economic loss. However, there are numerous ways to categorize risk. Some typical categories include:

Safety risk	Insurance risk
Health risk	Political risk
Environmental risk	Technological risk
Program risk	Portfolio risk
Security risk	Ecological risk
Financial risk	

In this chapter we focus on two types of risk: program risk and design risk.

Design for Reliability, First Edition. Edited by Dev Raheja, Louis J. Gullo.

Qualitative Risk Analysis

Qualitative risk analysis is the most widely used type of risk analysis, largely because it is quickest and simplest to perform. This approach uses terms such as *high, medium*, and *low* to characterize the risk. These levels of risk are determined by the combination of likelihood of occurrence and severity of loss in relative terms described in words rather than numerics. This form of risk analysis does not use actual hard data in the risk assessment process and as a result is subjective and relies heavily on the experience of the analyst.

Quantitative Risk Analysis

Quantitative risk analysis is used to estimate the probability of an undesirable event happening and to assess, in quantitative terms, the magnitude of the consequences using numerics to evaluate the risk. This approach is preferred, but it requires significant amounts of data, such as historical data or test results, to be able to accurately estimate the probability of occurrence and magnitude of the losses that may occur. The approach is complicated, time consuming, and costly. There is also uncertainty associated with the estimates. When data are limited or questionable, the risk assessment can be disputed, and communication of the risk becomes more difficult.

IMPORTANCE OF RISK MANAGEMENT

Managers and designers regularly handle risk in their jobs. Because of the ever-increasing complexity of projects and systems, risk of failures increases, and the consequences of failures can be quite substantial. Therefore, it is becoming more and more important for all risks to be identified, assessed, and mitigated. Risks must be managed more formally now than they were managed in the past.

Many benefits are achieved as a result of performing risk assessments. In addition to the obvious elimination or reduction of risk, assessments provide more subtle benefits:

- Improve the ability to meet or exceed schedule, budget, and technical performance requirements
- Increase depth of understanding of the product or system
- Identify and correct hazards and failures
- Develop design trade-offs among alternatives or options based on their relative risk

- Identify major contributors of risk
- Establish priorities for improving safety and reliability

Of course, there must be a balance between risk and benefit. The cost of eliminating all risks to all things at all times would be prohibitively expensive and time consuming. The program management and the design engineers must be able to assess the risks accurately and have enough information about the risks and rewards to make difficult decisions concerning how much risk is too much.

WHY MANY RISKS ARE OVERLOOKED

Risks are often overlooked for various reasons. In the following two sections we describe many of the most common reasons that risk is unforeseen and thus causes problems with the product or system being developed.

Requirements Generation and Analysis

The key to any successful program is to define the requirements properly. The premise of all good systems engineering is to define the system requirements clearly up front. If you don't know what you are building, you won't know how it will perform when it is built. The causes of many failures are incomplete, ambiguous, and poorly defined requirements. Early risk assessments help to focus the design efforts to eliminate risks associated with the product or system being developed. Missing or incomplete requirements result in design changes that are both costly and often increase the risk associated with a product.

Although it is obviously essential to clearly define the performance requirements of a product or system, it is equally important to define requirements for other characteristics of the product or system, such as modularity, reliability, safety, serviceability, logistics supportability, sustainability, human factors, maintainability, and fault tolerance. Other requirements of major importance are the physical and design interfaces. All interfaces, including mechanical and electrical interfaces, internal and external interfaces, user–hardware interfaces, and user–software interfaces, must be defined early. Interfaces are developed during the architecture design phase of a program. The better the interfaces are defined and standardized, the greater the probability of success during the integration and test phases of a program. The description of the environments in which the product or system will have to operate, be stored, and survive is also of critical importance in the requirements specification. Incomplete or inaccurate environmental specifications usually lead to system failures later in development or during customer use.

Two additional problems may arise if requirements analysis is inadequate early in the development process: requirements creep and requirements omission.

Requirements Creep

Requirements creep involves the continued generation and building of requirements throughout development, which is a typical cause of ever-increasing program costs. Although in one sense it is a good thing to add requirements and capability as a result of the unremitting, inexorable advancements in technology which lead to improvements in performance, it must be properly managed, even anticipated, if a program is to stay on budget. The U.S. presidential helicopter, the VH-71 program, is a prime example of a program's demise due to cost overruns resulting from requirements creep. After the contract was awarded, and after substantial designs were completed, more requirements were added, resulting in a doubling of estimated costs and serious delays in the program schedule. As a result, the program is planned for cancellation.

Requirements creep can also introduce problems in design. When new requirements are added, one must consider the effects of the new requirements on design. If safety and/or reliability analyses have been completed, they must be redone to capture the effects of the changes on the product or system. This is often not the case: Analyses are not revisited and the changed requirements result in additional risk entering the product or system.

Requirements Omitted

During the requirements development phase, requirements that are typically applied to similar systems are missing. These requirements may have been present in a boilerplate specification, but later removed due to restrictions on cost and schedule during program planning. Similar to requirements creep, there could be requirements that were missed in the requirements development stage. These requirements could have been missed due to ineffective requirements review and analysis.

Inadequate Cross-Functional Teams

Developing a good specification requires inputs and reviews from a diverse group of people. A cross-functional team composed of representatives of all stakeholders must be formed to help identify and agree to the requirements. The stakeholders include everyone with an interest or a stake in the results. Obviously, this should include the developer and the customer, but it should also include members from all the functional groups, such as reliability, safety, production, manufacturing, marketing, and testing. Leaving out key members of the team can only lead to incomplete requirements definition, requirements creep, and faulty products. Project management must integrate all the functions, the risks, and the people to ensure that a good product results from the development effort. All things associated with the development are interrelated and interconnected. It is important that management and the design team recognize this.

Risk Management Culture

It is imperative that a team culture of risk management be a driving force behind the engineering design effort. Making risk management culture a primary driver of engineering design starting from the informed approval of the requirements and permeating the entire design process will ensure robust product design. The entire team should be focused early on the development of the best and most complete requirements specification possible, and they must also have an additional focus on eliminating risk of all types, or reducing it to a minimum.

PROGRAM RISK

The process of project management always involves a certain amount of risk. The three major attributes that contribute to customer satisfaction in any project include delivering the right product or system, delivering it on time, and delivering it within the cost allocated. At the beginning of a new project, there are always uncertainties associated with these three factors. It is therefore important to manage the risk produced by these uncertainties in order to limit the negative consequences. We also, of course, want to maximize the positive possibilities presented by the project.

Cost, Schedule, and Performance

Generally speaking, there are three types of risk the project management is concerned with: cost, schedule, and performance (or technical) risk. Other more detailed breakdowns can be used, but these three risk types serve to encompass management's major concerns.

Cost is an obvious driver of risk concern. Management must minimize or eliminate cost risk. Cost risk relates to waste of resources, such as material, human capital, or energy.

Schedule risk is another driver of risk concern. Management must ensure that projects complete on time, get the product to market in a timely manner, or beat the competition to the market. Schedule risk must be managed, monitored, and controlled.

Performance or technical risk represents the third driver of risk concern. Technical risk may encompass numerous considerations, such as:

- Requirements
- Design
- Test and evaluation
- Technology
- Logistics: life-cycle supportability
- Production and manufacturing

Program Risk Management Process

The process of risk management can be handled in various ways, but the *Risk Management Guide for DoD Acquisition* [1] provides a good, generic approach to the risk management process. Figure 1 provides a snapshot of this process.

Risk management is concerned with decision making while considering uncertainty. It is a process wherein the risks are identified, ranked, assessed, documented, monitored, and mitigated. Managing risk entails identification of the risk; assessment of the risk, including the estimation of the likelihood of its occurrence and the consequences it presents; the cost–benefit trade-off; and the mitigation or acceptance of the risk. Risk management is used to evaluate events that could negatively affect the project. It is used to improve the likelihood of having a successful project.

Risk planning as part of an overall management planning effort is the process of documenting and implementing a comprehensive, proactive management strategy for identifying, handling, and continuously assessing risks.

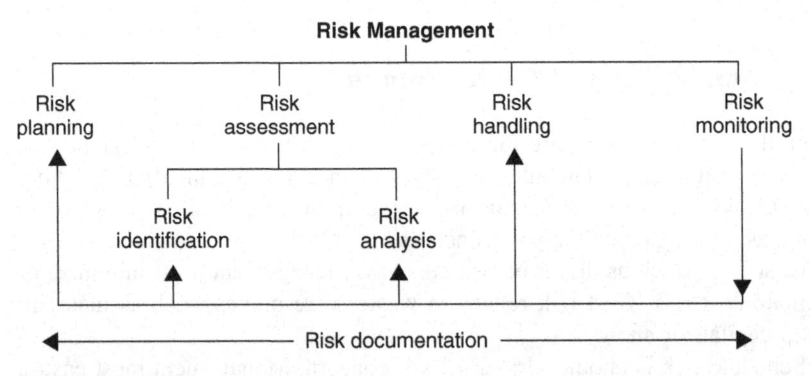

Figure 1 Risk management process.

Risk assessment is the process of identifying and analyzing project risks in an effort to increase the likelihood of meeting cost, schedule, and performance objectives.

Risk identification identifies the risk associated with each project activity and each technical area.

Risk analysis is the process of examining each risk identified to assign a probability of the undesirable event occurring and to determine the effects if the event occurs. This effort also entails ranking or prioritizing each risk in relation to the other risks, based on their probability of occurrence and the severity of their consequences. This process allows adequate resources to be brought to bear against the most important risks.

Risk handling is the process that identifies, evaluates, selects, and implements the approaches to mitigating the risks. This process establishes what is going to be done, who is going to do it, how much it's going to cost, and when it will be accomplished. Additional risks and risk-handling strategies may be added as the product development process matures.

Risk monitoring involves tracking the activities of risk handling. These activities are measured against metrics established by management and are evaluated regularly throughout the development process.

Risk documentation is critical to the risk management process. It entails regular status reports of the risk assessments and mitigation efforts, and recording the results of such efforts so that management has a constant appraisal of the status of all project risks. This documentation effort allows management to adjust resources as necessary to effectively mitigate the risks throughout the project duration.

DESIGN RISK

Another risk that we need to be concerned with is *design risk*. The preceding section dealt with program risk, the overall risks to a project or program. Technical risk was a subset of program risk, and design risk can be considered a subset of technical risk. However, since this book is focused on reliable and safe product design, we expand the topic of design risk in this section. Design risk can encompass several types of risk, such as engineering risk, safety risk, and risk of failure to perform the intended function. We have chosen to use the term *design risk* to encompass all these terms.

Current State of the Art of Design Risk Management

Risk is a concept that is common to both hardware and software products and products containing both. Concern over failures in products is a long-standing problem. The continuing growth of complexity in today's systems and products heightens the concern of risk. Because of increasing complexity, risks are increasing. Risks need to be identified, assessed, and managed in a more formal way than they have been in the past.

The engineer's job entails making technical decisions. Many of these decisions are made with limited or incomplete information. Incomplete information leads to uncertainty, and therefore risk is inherent in the engineering decision-making process. The traditional approach to reducing risk has been to design and regulate products or systems very conservatively. This may entail designing in large safety margins, including multiple safety barriers in the design, excessive quality control, and regular inspections. This conservative approach can lead to very expensive systems and still does not guarantee safety or success. The current trend is to use more formal methods of assessing risk through analysis.

Expression of Design Risk

Design risk is an expression of the possibility or impact of a mishap in terms of hazard severity and hazard probability. From the beginning of the design process the goal is to design to eliminate hazards and minimize the risks of failures, or to increase the probability of design success. If an identified hazard cannot be eliminated, the risk associated with it must be reduced to an acceptable level.

Risk is the product of consequences of a particular outcome (or range of outcomes) and the probability of its (their) occurrence. The most common way of quantifying risk is as the product of consequences of a particular outcome (or range of outcomes) and the probability of its (their) occurrence. This is expressed as

$$R = C \times P \text{ (consequence severity} \times \text{probability of occurrence)} \quad (1)$$

where the severity of the consequences of an undesirable event or outcome is an assessment of the seriousness of the effects of the potential event if it does occur. The probability of occurrence is the likelihood that a particular cause or failure will occur.

RISK ASSESSMENT

Risk assessment is the quantification of possible failures. To do a risk assessment, we need to know:

1. What can go wrong in the system?
2. How likely is the failure to happen?
3. What will be the consequence of the failure if it occurs?

All engineering design should involve the consideration of what will work and what might go wrong.

Risk combines the probability of failure with the consequences of failure. An essential factor in risk assessment is uncertainty. Uncertainty in design risk assessment weights the known with the unknown design failure modes, such as the case when the designer doesn't know exactly which failures may occur, how they will occur, or when or where they may occur. There is uncertainty associated with the three factors mentioned above: uncertainty as to what can go wrong, uncertainty in the likelihood as to a particular failure occurring, and uncertainty as to defining the consequences of a failure if it does occur. These uncertainties are due to a lack of knowledge about the risks. This lack of knowledge may be in the form of unasked questions, or unanswered questions, or questions with undesirable answers that are not documented or communicated. Our goal in reliable design is to minimize the uncertainty through increased knowledge, thus minimizing the uncertainty in our decisions on how to react to a risk, and minimizing the quantity of risks inherent in our designs.

Murphy's law states: "If something can go wrong, it will." Risk assessment provides a means to determine how likely something could go wrong, and what will happen if it does. Risk assessment provides the designer with a quantitative input to the design process and helps the designer to focus resources on the most significant problems (failures) in order to reduce exposure to risk.

RISK IDENTIFICATION

The first step to controlling risk is to identify those areas that present the risks. A designer may determine what can go wrong with the product or system being designed using numerous tools, such as failure modes, effects, and criticality analysis (FMECA). FMECA is useful to identify all design failure modes, or the designer may choose to use preliminary hazard analysis (PHA), which is used to determine the hazards that may be present in a product or system. PHA is used at the earliest stages of concept design to achieve "preliminary" results. As the design matures, PHA is expanded into other forms of hazard analyses, each being more detailed. PHA and other forms of hazard analysis are discussed in Chapter 17, and FMECA is described in several chapters. The reader may find more details of these techniques in the books by Raheja and Allocco [2] and Roland and Moriarty [3].

The key to the success of product design is the tracking of these failures and hazards throughout the design process to ensure that they are mitigated to acceptable levels. In addition to FMECA and hazard analysis, event tree analysis and fault tree analysis may be used to assess design risks. The fault tree analysis technique is also discussed in other chapters of the book.

Event Tree Analysis

Event tree analysis is an inductive analysis technique used to map the developments and consequences from a given initiating event. Event trees are especially useful for analyzing the effects of functioning or failing safety mechanisms and/or human reactions in response to the occurrence of an undesirable event. All the different possible sequences of events following the initiating event are identified systematically using the event tree.

Fault Tree Analysis

Fault tree analysis is a deductive analysis technique used to determine all paths that can lead to a particular, top event. Starting with the top, undesirable event, the analyst deducts what combination of next-lower-level failures or events can cause the top event. The analysis continues decomposing each of these failures or events to lower and lower levels until the lowest level of component failure is reached. At this point, all the possible paths to the top event can be calculated. Detailed quantitative analysis may also be conducted by applying failure rates to the lowest-level component failures to calculate the probability of the top, undesirable event occurring. This technique can also distinguish the critical failures and events that contribute most to the occurrence of the top event.

RISK ESTIMATION

As discussed previously, risk estimations are made by combining the probability of occurrence of the undesired event with a measure of the severity of the consequences. This combination of severity and probability is the risk assessment code. At the early stages of design, it is typical to assess the risk using an initial risk assessment code. This assessment is usually a worst-case assessment based on preliminary assessments made during the concept development stage of product design. Later, a "current" risk assessment code can be used to reflect the latest risk assessments. At the end of the development program, the current risk assessment code becomes the final risk assessment code.

Probability

Initially, a qualitative statement of estimated probability of occurrence is used to make the initial risk estimate. As the design progresses, and time and budget permits, the probability of occurrence can be refined to become more quantitative. This can be accomplished by using reliability analysis techniques to estimate product or system failure rates, or other analysis techniques, such as fault tree analysis or FMECA.

Consequences

Usually, the first step in assessing hazardous consequences is to make a qualitative ranking of the severity of the hazard. Much like estimating the probability of occurrence, the severity of consequences can be analyzed in greater depth as the design progresses, using more quantitative techniques, such as FMECA or event tree Analysis.

Cost of Mitigation

In addition to the probability of occurrence and the severity of the consequences, it is often necessary to consider the cost of mitigation. There are almost always budget limitations, and these must be taken into account during the process of deciding how much effort to expend to mitigate the risks and which ones to concentrate on. Obviously, the risks with the highest rank need the most attention, potentially to eliminate them, or to reduce the consequences if they do occur. Lower-ranking risks may be deemed acceptable due to cost constraints.

RISK EVALUATION

The next step is the evaluation of the risks to determine if further action is warranted or if the risk is acceptable. Risks must be evaluated to determine their significance and the urgency of their mitigation. Each risk is assigned a risk assessment code and as described in the risk assessment matrix, as shown in Figure 2. When using the matrix:

1. Use either the quantitative or qualitative descriptors of probability as appropriate for a given analysis.
2. Use either the individual item or fleet/industry description, depending on which description produces the more frequent probability level for a given analysis.
3. Probability level F is reserved for cases where the causation factor is eliminated or no longer present or it is impossible to lead to the mishap. No amount of documentation, training, warnings, cautions, personnel protective equipment, or other change can reduce a mishap probability level to F.

The risk assessment code is used to determine if further corrective action is required. Each risk's risk assessment code allows the relative ranking of the risks. The risks are typically coded using one of four choices that represent risk priority: high, serious, medium, or low. Further action can then be taken on the most important risks first. The ranking of the hazard into these groups

Specific Individual Item	Fleetor Industry	Probability / Severity	Catastrophic (1)	Critical (2)	Marginal (3)	Negligible (4)
			Could result in death, permanent total disability, loss exceeding $1M, or irreversible severe environmental damage that violates law or regulation.	Could result in permanent partial disability, injuries or occupational illness that may result in hospitalization of atleast three personnel, loss exceeding $200K but less than $1M, orreversible environmental damage causing aviolation of law or regulation.	Could result in injury or occupational illness resulting in one or more lost work days(s), loss exceeding $20K but less than $200K, or mitigatible environmental damage without violation of law or regulation where restoration activities can be accomplished.	Could result ininjury or illness not resulting in a lost work day, loss exceeding $2K but less than $10K, or minimal environmental damage not violating law or regulation.
Likely to occur often in the life on an item; with a probability of occurrence greater than10^1 in that life.	Continuously experienced	Frequent (A)	High	High	Serious	Medium
Will occur several times in the life of an item; with aprobability of occurrence less 10^{-1} than but greater than 10^2 in that life.	Will occur frequently	Probable (B)	High	High	Serious	Medium
Likely to occur sometime inthe life of an item; with a probability of occurrence less than 10^2 but greater than10^3 in that life.	Will occur several times	Occasional (C)	High	Serious	Medium	Low
Unlikely, but possible to occur in the life of an item; with a probability of occurrence less than10^3 but greater than10^6 in that life.	Unlikely but can reasonably be expected to occur	Remote (D)	Serious	Medium	Medium	Low
So unlikely, it can be assume occurrence may not be experienced during a defined interval in the life of an item; with a probability of occurr ence of less than 10^6 in that life.	Unlikely to occur, but possible	Improbable (E)	Medium	Medium	Medium	Low
Incapable of occurrence in the life of an item.This category is used when potential hazards are identified and later eliminated.	Incapable of occurrence within the life of an item. Thiscategory is used when potential hazards are identified and later eliminated.	Eliminated (F)	Eliminated			

Figure 2 Risk assessment matrix.

Table 1 Risk Levels

Risk level	Risk assessment code	Guidance	Decision authority
High	1A, 1B, 1C, 2A, 2B	Unacceptable	Management
Serious	1D, 2C, 3A, 3B	Undesirable	Program manager
Medium	1E, 2D, 2E, 3C, 3D, 3E, 4A, 4B	Acceptable	Program manager or safety manager
Low	4C, 4D, 4E	Acceptable (without higher-level review)	Safety team

determines if further action is necessary and by what authority the hazard can be accepted and closed (see Table 1 for risk levels).

RISK MITIGATION

Risk mitigation uses design measures to reduce the probability of occurrence of a failure or undesired event and/or by reducing the consequences that result if the failure or undesired event occurs. The best solution is always to eliminate the risk by designing it out of the system or product. The first step in mitigation of risk is the identification of risk mitigation measures. The designer must identify potential risk mitigation alternatives and their expected effectiveness. Risk mitigation is an iterative process that culminates in either complete elimination of the risk or in a residual risk when the risk has been reduced to a level acceptable to the decision-making authority.

The order of precedence for mitigating identified risks is provided in MIL-STD-882 [4]:

1. *Eliminate hazards through design selection*. If unable to eliminate an identified hazard, reduce the associated risk to an acceptable level through design selection.
2. *Incorporate safety devices*. If unable to eliminate the hazard through design selection, reduce the risk to an acceptable level using protective safety features or devices.
3. *Provide warning devices*. If safety devices do not adequately lower the risk of the hazard, include a detection and warning system to alert personnel to the particular hazard.
4. *Develop procedures and training*. Where it is impractical to eliminate hazards through design selection or to reduce the associated risk to an acceptable level with safety and warning devices, special procedures and training must be developed. This is the least effective and thus least desirable way to control risks.

Robust design is the best way to mitigate product or system risk.

RISK COMMUNICATION

Risk communication is the exchange of information between interested parties concerning the nature, scale, importance, disposition, and control of risk. Risk communication can, and should, occur at all stages of the risk assessment process. On a macro scale, interested parties could be government agencies, companies, unions, individuals, communities, and the media. Although ultimately these external parties may need to be informed of the risk associated with a product, the product's maker is more interested in the internal risk communication process that must take place during the design process.

For any product design effort, the program/project manager is responsible for the communication, acceptance, and tracking of hazards and residual risk. The program manager must communicate known hazards and associated risks of a system to all interested parties. As changes are introduced into the system, the program manager must update the risk assessments. The program manager is also responsible for informing system designers about the program manager's expectations for handling newly discovered hazards. The program manager will evaluate new hazards and the resulting residual risk and either recommend further action to mitigate the hazards or formally document the acceptance of these hazards and residual risk. The program manager must evaluate the hazards and associated residual risk in close consultation and coordination with the ultimate end user, to assure that the context of the user requirements, potential mission capability, and operational environment are addressed adequately. Providing documentation of the hazard and risk acceptance to interested parties facilitates communication. Any residual risks and hazards must be communicated to system test efforts for verification.

The designer is responsible for communicating information to the program manager on system hazards and residual risk, including any unusual consequences and costs associated with hazard mitigation. After attempting to eliminate or mitigate system hazards, the designer should formally document and notify the program manager of all hazards breaching thresholds set in the safety design criteria. At the same time, the designer will also communicate the system residual risk.

Key to the successful risk communication process are the lines of communication within the program organization and with the system safety organization and associated engineering organizations. Interfaces must be established between system safety and other functional elements of the program. Also important to the risk communication process is the establishment of the authority for resolution of identified hazards.

Another critical functional area within the project organization that plays a major role in risk resolution and communication is the test organization. Risk mitigations that have been implemented in the design must be tested for effectiveness during the system test process. Residual risk and associated hazards must be communicated to the system test efforts for verification. Communication with a test organization is a two-way street. Lines of communication between the test organization and the rest of the program team must remain open and candid. All identified hazards, safety discrepancies, and product failures found during testing must, as a minimum, be communicated by the testing organization to design engineering and program management.

RISK AND COMPETITIVENESS

Program and design risks can affect competitiveness. There is a risk that the final cost will exceed the planned or expected cost. The introduction of a product into the marketplace can result in profit loss or gain. The goal is to make design decisions that maximize the profit gain and minimize the loss. There is a risk that the planned introduction of the product being designed will fall behind schedule and will lose out to the competition. There is a risk that the performance of the product will fall short of expectations, thus losing out to a competitor. Every engineering decision comes with risk and has the possibility of loss associated with it.

RISK MANAGEMENT IN THE CHANGE PROCESS

With change comes risk. A major area where risk, either program risk or design risk, can enter a project is when changes are made. Changes in a program often bring with them changes to schedules, cost, or technology. All design changes, no matter how small or insignificant, must be evaluated for the added risk they may pose to the program and for any possible negative effects on the design. A design change may bring with it technical risk and additional risk to a safe, reliable design. The added risk may not even have been considered earlier in the product's risk assessment.

CONFIGURATION MANAGEMENT

It is obviously best never to make changes, especially late in the development process. As emphasized previously, the complete specification of requirements early in the development process is critical to the minimization of changes and the success of the program. However, change is inevitable. A robust configuration management process is needed to ensure that changes are made in an orderly fashion and to minimize additional risk being added by changes.

Definition of Configuration Management

Configuration management is defined by MIL-HDBK-61 [5] as "a process for establishing and maintaining consistency of a product's performance, functional and physical attributes with its requirements, design, and operational information throughout its life." The configuration management process is used to control system products, processes, and related documentation. The configuration management effort includes identifying, documenting, and verifying the functional and physical characteristics of an item, recording the configuration of an item, and controlling changes to an item and its documentation. It should provide a complete audit trail of decisions and design modifications. The reader can find more details on the implementation of a configuration management process in MIL-HDBK-61 [5].

Benefits, Risks, and Cost Impact of Configuration Management

Configuration management (CM) provides knowledge of the correct current configuration of the product or system being developed. The CM process manages necessary changes efficiently, ensuring that all impacts to operation and support are addressed. The main benefits of the CM process are summarized as follows:

- Product attributes are defined. Both the developer and the user have a common basis for understanding and use of the product.
- Product configuration is documented and a known basis for making changes is established.
- Proposed changes are identified and evaluated for impact prior to making change decisions. Surprises are avoided.
- Change activity is managed using a defined process.
- A high level of confidence in the product information is established.

In the absence of CM or where CM is ineffective, there may be equipment failures due to incorrect part installation or replacement; schedule delays and increased cost due to unanticipated changes; operational delays due to mismatches with support assets; maintenance problems, downtime, and increased maintenance cost due to inconsistencies between equipment and its maintenance instructions; and numerous other circumstances that decrease operational effectiveness and add cost. The intent of CM is to control changes to avoid cost and minimize risk. Those who consider the small investment in the CM process a cost driver may not be considering the compensating benefits of CM and may be ignoring or underestimating the cost, schedule, and technical risk of an inadequate or delayed CM process.

Handling Risk during the Change Process

As in the front-end analysis effort that leads to good, complete definition of requirements, it is important to have representatives from all the functional disciplines involved in the change management process. This inclusive approach helps to ensure that all the risks associated with each change proposed will be thoroughly evaluated for the total effect on the product or system.

REFERENCES

[1] "Risk Management Guide for DoD Acquistions", 2nd ed., May 1999.
[2] Raheja, D. G., and Allocco, M., *Assurance Technologies Principles and Practices: A Product, Process, and System Safety Perspective*, 2nd ed., Wiley, Hoboken, NJ, 2006.
[3] Roland, H. R., and Moriarty, B., *System Safety Engineering and Management*, Wiley, Hoboken, NJ, 1990.
[4] *Standard Practice for System Safety*, MIL-STD-882D, U.S. Department of Defense, Washington, DC, Feb. 2000.
[5] *Configuration Management Guidance*, MIL-HDBK-61, U.S. Department of Defense, Washington, DC, Feb. 2001.

Chapter 17

Integrating Design for Reliability with Design for Safety

Brian Moriarty

INTRODUCTION

Design for safety is a major effort to assure that all known hazards have been considered for a product being designed, and that the hazards are mitigated or acceptable for system operation. In this chapter we address the focus points necessary to assure that a well-designed "acceptable hazard" system will be built. Reliability is a key feature in obtaining the total product base to examine hazards. Previous chapters emphasized the importance of knowing the reliability of the system, the subsystem, and the components of all areas of a product. The safety team utilizes this information in the course of examining a design to insert the reliability data and obtain the risk assessment codes for each hazard found in the design. Design for safety and design for reliability are inseparable. The design project team must work together in the directly compatible use of the information, to reach the conclusion that the design can be used reliably with the hazards defined as acceptable hazards.

The goals of system safety are to have complete knowledge of a system from its original generic design to the final design, with an examination of the hazards that exist, including human errors that may lead to unsafe consequences [1]. Documenting the importance level of each design analysis for safety is a key factor in the history of building a design, and records are

Design for Reliability, First Edition. Edited by Dev Raheja, Louis J. Gullo.

kept at each step of the building process in which safety has been analyzed. The use of a previous "hierarchy system" from which a revised product is made begins with an examination of the previous product and the safety status of that system [2]. Previous systems, having gone through acquisition to final operational field use, assist greatly in creating a knowledge base of the enhanced system that will serve as a revision update to the hierarchy system.

START OF SAFETY DESIGN

Safety design includes examining boundaries in the architectural design of the product. Initial functional system architecture of the product must be defined. Second, initial generic specifications for the product must be defined to start the first safety hazard analysis, which would be part of the total hazard analysis work. These boundaries must be included in the requirements for design for reliability to design for safety [4]. The list of information to obtain should include:

- System architecture of the previous acquisition product
- System architecture of the enhanced system
- System specification of the acquisition product
- Functional system specification of the enhanced system
- Safety hazard analysis data of the acquisition product
- Reported safety problems with the acquisition product
- Safety investigation reports of hazard problems with the acquisition system

The basic equipment that will be designed for the product must be defined in its beginning stage with boundary levels of the product that are known from the generic listing at the first development phase of the product. This would include the:

- Worst-case environment in which the system would be used
- Personnel interfaces of the product with other products and systems
- Procedures necessary for safety operation of the product under all usage conditions
- Facilities that are necessary for product use
- Training of operators, maintainers, and users of the product

The interfacing between environment, personnel (user, operator, and maintainer), procedures of operation and maintenance, facilities for equipment, and training are key items that are under the umbrella of interrelationship. One factor that is degraded, such as environment with a high temperature above specification, can affect the other portions of the design. A distinct understanding of the interfaces of each of these factors is a part of the safety evaluation that is required as the product is developed.

RELIABILITY IN SYSTEM SAFETY DESIGN

Research of reliability data related to product design is a major factor due to development of hazard analysis. Evaluation of the likelihood of failures that are associated with defined hazards becomes a leading factor in further definition of the risk level of a particular hazard. Methods by which reliability can be obtained are discussed elsewhere in the book. The hazard cause will have a corresponding failure mode that will provide information that can be used in the further definition of risk level.

Tolerance factors (low to high values for reliability with confidence levels) in the reliability of a component, subsystem, system, ability to follow a procedure, and so on, must also be defined. In most hazard analysis forms the worst-case application of reliability tolerance is used, representing the hazard level in the scenario of the system proposed. In addition, the life cycle of the component, product, or system must be known. The life cycle and applications that should be made when conducting worst-case reliability have been discussed in other chapters. This will include the periodic maintainability service that should be performed to maintain a steady failure rate which would be used in the likelihood application to a defined hazard.

SAFETY ANALYSIS TECHNIQUES

The safety design involves a series of analysis techniques that are timelined into the product development. Product development begins with a research effort to define various development methods by performing several of the following analysis techniques [6].

Comparative safety analysis (CSA) examines proposed designs to understand the hazards involved and then evaluates the options for design to minimize hazard presence. The result is a recommendation for the safety design based on the presence of potential hazards that could be in a more detailed design that would be authorized.

Preliminary hazard analysis (PHA) is performed on the first generic design selected using the system architecture and functional specification documentation. An initial list of hazards with an evaluation of the risk assessment code is developed. A hazard worksheet is the data listing of hazards from the source documents available and the safety designer's evaluation team, which originates the worksheet. Hazard worksheet definitions are shown in Table 1.

Table 1 Hazard Worksheet Definitions

(1) Hazard ID	(2) Hazard description	(3) Causes	(4) System state	(5) Possible effects	(6) Severity/ rationale	(7) Existing controls or requirements	(8) Likelihood/ rationale	(9) Current or initial risk	(10) Recommended safety controls or requirements	(11) Predicted residual risk
Name of hazard: title with number for each hazard examined in the system	Safety hazard is any real componential condition that can cause (1) injury, illness, or death to people; (2) damage to or loss of system; or (3) damage to the environment	An event that results in a hazard or failure that causes a hazard; causes can occur independently or in combination	Definition of credible conditions in which a system exists	The worst-case effects of the hazard occurrence	Listing of Severity of hazard with rationale for the selected severity	List of the controls of the defined system that eliminate or minimize the occurrence of the hazard	Estimation for each hazard of how often the effects of the hazard will occur; this is related to the worst-case system state and possible effects	Combined severity and likelihood that are identified as the risk assessment code (RAC) will be identified as a high-, medium-, or low-level hazard	Identification of mitigation requirements required to lower the current or initial risk RAC; this list represents added requirements for the system which are then reevaluated for the residual risk predicted	Reevaluation of risk in an RAC definition table with selection of mitigation safety controls and requirements placed into the defined configured system

Subsystem hazard analysis (SSHA) performs a safety risk assessment of a system's subsystems and components as a greater level than provided in PHA. Specific purposes are (1) to verify subsystem compliance with system and safety requirements, (2) to identify previously unidentified hazards associated with the subsystem, (3) to assess the risk of the subsystem design, and (4) to consider human factors, functional and component failures, and functional relationships between components comprising the subsystem, including software.

System hazard analysis (SHA) is developed as a detailed safety risk assessment of a system examining (1) the interfaces of that system with other defined systems, and (2) the interfaces between the subsystems that comprise the total system. Specific purposes are the same those identified for subsystem hazard analysis.

Operating and support hazard analysis (O&SHA) examines the safety risk assessment of a system's operational and support procedures. This involves (1) evaluating all operating and support procedures for the human-to-equipment procedure operation, (2) identifying hazards with procedures, (3) examining human factors and critical human errors in normal and emergency procedures and support tasks, (4) assessing the risk of existing procedurals and controls, and (5) developing alternative controls and/or procedures to eliminate or control identified hazards.

Fault tree analysis (FTA) is a deductive method of analysis that is used exclusively as a qualitative analysis or, if desired, can be expanded to a quantitative analysis. The primary item that fails will have a series of AND and OR gates to show graphically the combination of these gates leading to each of the failures of an item. Each part of FTA asks the question: What creates the failure of the item listed? The purpose is to build a logic tree that will identify these causes. Second, probability figures for the failures can be assigned with Boolean algebra used to estimate mathematically the total probably of failure or malfunction for the primary item that is at the top of the fault tree.

A system safety assessment report (SSAR) is a summation of the hazard analysis data found in performing the previous hazard analyses. Specific purposes include (1) collecting the results of safety risk management controls; (2) identifying all safety features of hardware, software, and system design; and (3) identifying procedural, human factors, hardware, software and software-related hazards that have been identified in the program to date, and assessing system readiness based on the cumulative safety risk so that final testing and operation of the system can proceed.

More detailed information on these related hazard analysis may be found in the references. Table 1 presents a standard hazard worksheet with explanations of the data in the columns.

Definitions of the severity of a hazard are assigned with the criteria shown in Table 2. The safety designer must evaluate the result of the hazard in examination of the product.

Table 2 Severity Definitions

Description	Category	Environmental, safety, and health result criteria
Catastrophic	I	Could result in death, permanent total disability, loss exceeding $1 million, or irreversible severe environmental damage that violates law or regulation.
Critical	II	Could result in permanent partial disability, injuries, or occupational illness that may result in hospitalization of at least three personnel, loss exceeding $200,000 but less than $1 million, or reversible environmental damage, causing a violation of law or regulation.
Marginal	III	Could result in injury or occupational illness resulting in one or more lost workdays, loss exceeding $20,000 but less than $200,000, or mitigation environmental damage without a violation of law or regulation where restoration activities can be accomplished.
Negligible	IV	Could result in injury or illness not resulting in a lost workday, loss exceeding $2000 but less than $10,000, or minimal environmental damage not violating law or regulation.

The safety designer must understand the probability definitions used in examining the product being developed. Each hazard found and documented in the hazard worksheet is assigned a probability level, as outlined in Table 3. This is later combined with the severity level and united in the risk assessment code, which represents the risk level associated with the hazard being examined, as shown in Figure 1.

The safety order of precedence to control hazards is one of the fundamental principles of system safety in eliminating, controlling, or mitigating a hazard. Table 4 defines the order in which action is taken to control a hazard. The use of reliability data from failure modes and effects analysis (FMEA) is used to find any single-point or common-cause failure [3]. If *no* single-point or common-cause failures exist, the likelihood can be reduced to a lower level by the definitions in Table 4.

The major effort is made to eliminate the risk designated as priority 1. Designing to minimize a risk to an acceptable level will always be the goals of the safety designer. The order of action taken is shown clearly in Table 4 and is reviewed by program management to assure that action is taken in following the precedence. In many cases there will be a combination of several of the actions to have combinations of priorities 1, 2, 3, and 4 to assure that all aspects of design safety control are placed into the overall design of the product.

Table 3 Probability Definitions

Description	Level	Specific individual item	Fleet or inventory
Frequent	A	Likely to occur often in the life of an item, with a probability of occurrence greater than 10^{-1} in that life.	Experienced continuously.
Probable	B	Likely to occur several times in the life of an item, with a probability of occurrence less than 10^{-1} but greater than 10^{-2} in that life.	Will occur frequently.
Occasional	C	Possible to occur some time in the life of an item, with a probability of occurrence less than 10^{-2} but greater than 10^{-3} in that life.	Will occur several times.
Remote	D	Unlikely but possible to occur in the life of an item, with a probability of occurrence less than 10^{-3} but greater than 10^{-6} in that life.	Unlikely, but can reasonably be expected to occur.
Improbable	E	So unlikely that it can be assumed that occurrence may not be experienced, with a probability of occurrence less than 10^{-6} in that life.	Unlikely to occur, but possible.

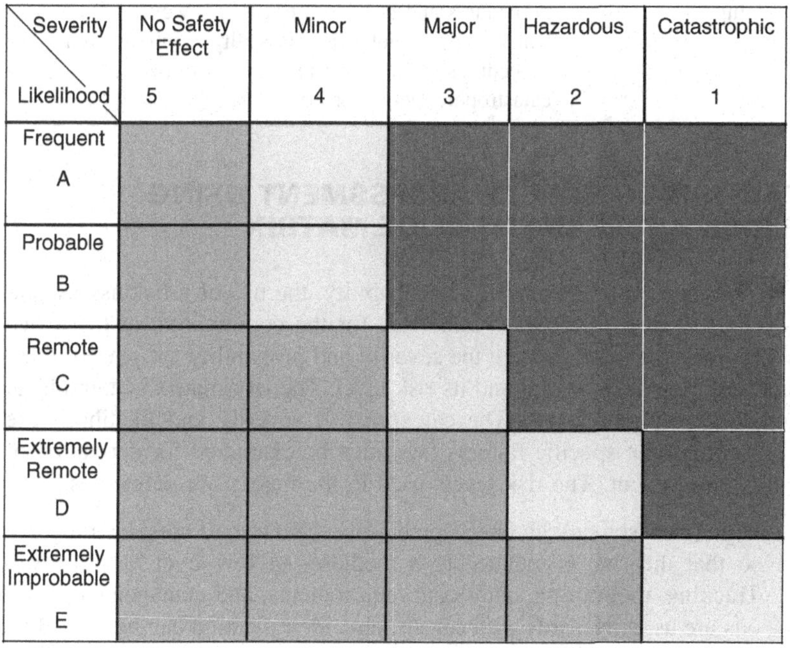

Severity / Likelihood	No Safety Effect 5	Minor 4	Major 3	Hazardous 2	Catastrophic 1
Frequent A					
Probable B					
Remote C					
Extremely Remote D					
Extremely Improbable E					

Figure 1 Risk assessment code matrix.

Table 4 Safety Order of Precedence

Description	Priority	Definition
Design for minimum risk	1	Design to eliminate risks. If the defined risk cannot be eliminated, reduce it to an acceptable level through design selection.
Incorporate safety devices	2	If defined risks cannot be eliminated through design selection, reduce the risk by the use of fixed, automatic, or other safety design features or devices. Provisions shall be made for periodic functional checks of safety devices.
Provide warning devices	3	When neither design nor safety devices can effectively eliminate identified risks or reduce risk adequately, devices are used to detect the condition and to produce an adequate warning signal. Warning signals and their use are designed to minimize the likelihood of inappropriate human reaction and response. Warning signals and placards are provided to alert operational and support personnel of suck risks as exposure to high voltage and heavy objects.
Develop procedures and training	4	Where it is impractical to eliminate risks through design selection or specific safety and warning devise, procedures and training are used. However, concurrence of authority is usually required when procedures and training are applied to reduce catastrophic, major, or critical safety.

ESTABLISHING SAFETY ASSESSMENT USING THE RISK ASSESSMENT CODE MATRIX

With the definitions for severity and probability, the use of a risk assessment code (RAC) is used to categorize the risk for the combination of these two areas. Figure 1 is a matrix where the severity and probability are placed in the box that represents the hazard and its risk level. The risk matrix is a graphical method to define risk levels. The categories of severity and likelihood are keys to defining the specific failures that must be examined for derivation of reliability assignment. The risk levels used in the matrix are defined as:

1. High (dark screen): an unacceptable risk. The hazard must be mitigated so that the risk is reduced to a medium- or low-level hazard level. Tracking, monitoring, additional requirements, and management methods are used. Hazards with catastrophic identification can be caused by single-point failures or events, common-cause failures or undetectable events in combination with single-point or common-cause events.

2. Medium (lightest screen): an acceptable risk. This meets a minimum acceptable safety objective. Additional requirements and changes can still be implemented to improve the hazard to a lower-level risk assessment code. Continuous tracking monitoring and management checks are still required.
3. Low (medium screen): an acceptable risk with no additional requirements of mitigation. However, additional requirements can be added. Continuous tracking, monitoring, and management checks are always made to assure that this low-level hazard does not change.

DESIGN AND DEVELOPMENT PROCESS FOR DETAILED SAFETY DESIGN

All subsystems of total product must be defined to conduct a hazard analysis of the total system. Also, interfaces must be defined in the initial definition of the subsystems to clarify the flow of data that will exist between the subsystems. After knowledge of the internal interfaces, the external interfaces will be defined so that all areas of access (to and from the system) will be known [5]. Finally, operational use of the system must be defined so that the method of operation and maintenance for the system are known.

For any system developed, users and their ability to operate the defined system become very important for maintaining a safe system. A full comprehensive list of users performing all parts of the defined system must be known. Operational errors can occur with the user (or maintainer of a system/subsystem) that represent a major high-level hazard to the equipment as well as to the user. With this information known, the detailed hazard worksheets can be developed that analyze the hazards associated with a product (or system) to determine the risk level (high, medium, or low).

VERIFICATION OF DESIGN FOR SAFETY INCLUDES RELIABILITY

In the timetable of product development, testing will be required to prove that the safety control requirements are properly met. Testing programs require a test plan, test procedures, and a test report to record all results of the test. As the safety control requirements are defined to lower subsystem levels from the functional specification documents, this list must be assembled to assure that complete testing is performed. The designer for safety *MUST* be involved in the testing program to witness and review the testing results. There must be Problem Reports (PRs) that have been written against the safety requirements that fail their test. Follow-on work must be performed by the safety designers to resolve the problem report and reach a conclusion of the change required with a retest to be performed, known as regression testing. Follow-through work in reaching a "passed" status of the safety control requirement is the goal that must be achieved.

Testing of safety control requirements can be performed by:

- Testing with the product in operation with the conditions of operation defined by the overall operational specification manual
- Analysis performed by examination of acceptable methods of complete analysis for proof of the requirement
- Inspection of the product (or its parts) to show that it meets the requirement of operation by the specification or standard
- Demonstration that the product operation performs its operation correctly in terms of the required operation method defined in the accepted product manual
- Special tests performed for the product under extreme environmental operation for proof of acceptance

A review of design documentation to reach an accepted hazard of a final reliable product must be performed. Testing methods for safety must show the ability to achieve reliability of the RAC (severity and likelihood) by (1) qualitative reliability data, (2) quantitative reliability data, or (3) the life cycle of product data. The test performed must show proof of the integrity of the hazard RAC that was found in the design of the product. If a test results in the product not meeting its RAC-established hazard risk level, the RAC *must* be reexamined. In the situation where the RAC is a lower level of risk than found previously in the analysis, the test results will probably be accepted. However, when the RAC changes results in a higher level of hazard definition, action must be taken to reexamine the product properly for a change in order to bring the product to an acceptable hazard level. The designer for safety is responsible for assuring that action is taken and resolution of change is reached before a change can be made to the product and a retest can be performed. However, for that change to take place, the reexamination will include (1) brainstorming on the new hazards from the change proposed, and (2) risk mitigation of new hazards.

EXAMPLES OF DESIGN FOR SAFETY WITH RELIABILITY DATA

The following examples show the method of hazard analysis that needs to be performed for a specific product. The product is described followed by the purpose of the product and the basic functional requirements of the product. The following three examples are used: (1) use of a first product with no previous design product, (2) development of a product that has a large number of interfaces, and (3) development of a design added to the present product or system.

Use of First Product with No Previous Design Product

Product: infant crib (age: birth to 1 year)
Purpose of product: To provide an area for an infant to be in a horizontal position (either awake or sleeping).
Basic functional requirements: The crib has (1) an open wooden fence with vertical bar openings on two sides. (a barrier exists at the head and foot of the infants position), (2) a rectanglar design of a crib with two fences and two barriers, (3) a gate in the fence to open vertically when a parent is to open the gate using a switch, and (4) a low-level mattress in the crib for an infant to lie comfortably in a horizontal position.

(1) Hazard ID	(2) Hazard description	(3) Causes	(4) System state	(5) Possible effects	(6) Severity/ rationale	(7) Existing controls or requirements	(8) Likelihood/ rationale	(9) Current or initial risk	(10) Recommended safety controls or requirements	(11) Predicted residual risk
Crib-1	An infant pulls a mattress over his or her head.	The infant is able to pull a mattress because no tie-down of the mattress to the crib is present.	Use of crib for infant anytime during the 24-hour day.	Any time that the infant can pull the mattress over his or her head, the potential effect on the infant's breathing can be unacceptable.	1: catastrophic fatality due to suffocation of the infant.	**1.** Mattress with no tie-down to crib. **2.** No other method to prevent mattress from changing position.	D: extremely remote; difficult for infant to pull a mattress his or her over head.	1D: high-level hazard; unacceptable.	**1.** Place tie-down lines on the mattress and connect them to the crib. **2.** Develop a procedure in the crib manual to alert the crib owner to the necessity to tie down the mattress. **3.** Add an alarm system bell that will come on if the mattress tie-down lines are opened.	3B: low-level hazard; acceptable.

Development of Product That Has Large Number of Interfaces

Product: refrigerator

Purpose of product: To maintain food in a temperature-controlled refrigerator.

Basic functional requirements: A refrigerator provides a controlled lower temperature for food enclosed in its enclosed framed box. House power controls the refrigerator temperature using Freon tubing. Cold water is provided to the refrigerator with water tubing. A separate power outlet is provided for the refrigerator in the circuit breaker box.

(1) Hazard ID	(2) Hazard description	(3) Causes	(4) System state	(5) Possible effects	(6) Severity/ rationale	(7) Existing controls or requirements	(8) Likelihood/ rationale	(9) Current or initial risk	(10) Recommended safety controls or requirements	(11) Predicted residual risk
Reg-1	Refrigerator fails to operate.	House power fails; temperature control switch fails to operate; freon tubing fails; power outlet fails; water tubing breaks and shorts power, causing circuit breaker to open.	Use of refrigerator cannot keep food at designated temperature.	Contaminated food from unregulated temperature.	3: moderate.	Provide additional separate house power outlet to reconnect refrigerator to other operating power outlet; reconnect circuit breaker if power disconnects from power short.	C: low.	3C: moderate-level hazard.	Add backup power for refrigerator; place additional instructions in operator's manuals for user to gain more information on refrigerator hazards.	4E: low-level hazard; acceptable.

Development of Design Using Past Historical Reliability Data

Product: uninterrupted primary power supply (UPS) for function board in nuclear power plant
Purpose of product: To provide an additional power supply to back up the primary power supply in case of a loss of primary power.
Basic functional requirements: UPS is to be wired with a double-sided switch to connect when the primary power supply fails.

(1) Hazard ID	(2) Hazard description	(3) Causes	(4) System state	(5) Possible effects	(6) Severity/ rationale	(7) Existing controls or requirements	(8) Likelihood/ rationale	(9) Current or initial risk	(10) Recommended safety controls or requirements	(11) Predicted residual risk
UPS-1	UPS cannot function when required to operate.	Switch fails in change to provide power when primary power fails; UPS is not operational, due to incorrect wiring; UPS has a failed starter.	Use of UPS cannot provide power to a functional board.	Functional board in a nuclear power plant fails, causing loss of power to control water temperature; water temperature rises, causing nuclear plant explosion.	1: catastrophic.	UPS functional board power switch is checked daily.	D: extremely remote.	1D: high-level hazard; unacceptable.	Add signal generator for backup to UPS if UPS fails to take over basic power; add continuous check of UPS switch operability; add alarm alert machinery to functional board operator to inform that UPS will not work; add additional operator manual data for additional requirements; add training to nuclear power plant personnel concerning loss of power for emergency action.	1E: medium acceptable.

FINAL THOUGHTS

Safety designers must always consider the reliability of the products for which they are responsible. Probability information is a key to the hazard analysis that must be carried out on the product. Knowledge of the hazards associated with the product must be performed so that the risk of the hazard can be evaluated properly. It is recognized that there will always be hazards in a product. Electronic equipment has power that must be evaluated for the use of operators, maintainers, and users. The effort in design for safety is to recognize all hazards, including human errors, evaluate them properly, and assure that the hazards are acceptable.

Safety and reliability are inseparable. The standard risk assessment code assigns a RAC number to the combination of severity and likelihood that places the evaluation in the low, medium, or high hazard level. Based on this evaluation, the final step is to determine if additional requirements are needed to mitigate the hazard to an acceptable hazard level so that the product can be used. Without the reliability data, the total evaluation of the hazard cannot be completed. The hazard level associated with the failure resulting in the hazard occurrence *MUST* be defined in the evaluation the safety designer performs on the product.

The examples provided illustrate the method of developing a hazard worksheet that will document the product and the data needed for total evaluation of product hazards. Detail in listing the hazard data is a major factor in this evaluation. Peer review of the hazard data is normally performed as an independent review to assure that the data are correct and the hazard risk level is displayed correctly. Definitions of the worst credible event occurring from the hazard must be shown with the associated data for the exact hazard-level information in the design for safety.

REFERENCES

[1] *Department of Defense Standard Practice System Safety*, MIL-STD-882D, U.S. Department of Defense, Washington, DC, Jan. 1993.
[2] *Federal Aviation Administration System Safety Handbook*, FAA, Washington, DC, Dec. 2000.
[3] Roland, H., and Moriarty, B., *System Safety Engineering and Management*, 2nd ed., Wiley, New York, 1990.
[4] Raheja, D. G., *Product Assurance Technologies: Principles and Practices*, Wiley, New York, 1995.
[5] Raheja, D., and Allocco, M., *Assurance Technologies: Principles and Practices, A Product, Process, and System Safety Perspective*, 2nd ed., Wiley, Hoboken, NJ, 2006.
[6] *System Safety Analysis Handbook: Source Book for Safety Practitioners*, System Safety Society, Unionville, UA, 2005.

Chapter 18

Organizational Reliability Capability Assessment

Louis J. Gullo

INTRODUCTION

Systems integrators (SIs), original equipment manufacturers (OEMs), product designers (PDs), and product manufacturers (PMs) tightly manage their supply chains across the various manufacturing sectors for devices, materials, and equipment, to ensure that they satisfy their customers' requirements and needs. Those SIs, OEMs, PDs, and PMs who want to ensure continuous successful delivery of highly reliable systems and products to their customers will determine the capability of their suppliers within their supply chain. The organizational reliability capability of the supplier base must be assessed and understood by SIs, OEMs, PDs, and PMs if they wish to achieve success and maintain a competitive advantage through the design, manufacture, and sustainment of reliable systems and products.

In this chapter we provide guidance on how to assess an organization's or supplier's core competencies related to the organization's reliability capability. We focus on the purpose and objectives of the new IEEE standard for assessing organizational reliability capability, IEEE 1624–2008, which was published in February 2009. IEEE 1624–2008 is used to describe how the reliability capability of an organizational entity is determined by assessing eight key reliability practices and associated metrics. We explain the benefits of IEEE 1624–2008 in assessing the capability of an organization to deliver a reliable

Design for Reliability, First Edition. Edited by Dev Raheja, Louis J. Gullo.

product, which is termed organizational reliability capability. A reliability analyst determines an organization's reliability capability by assessing eight key reliability practices:

1. Reliability requirements and planning
2. Training and development
3. Reliability analysis
4. Reliability testing
5. Supply chain management
6. Failure data tracking and failure analysis
7. Verification and validation
8. Reliability improvements

These key practices encompass all aspects of critical design processes and build operations in the developer's or manufacturer's organization from a product reliability perspective.

A direct correlation exists between an organization's capability maturity and an organization's capability to develop mature and reliable products. A mature organization is capable of repeatedly designing and producing mature products. To develop a mature organization with the capability to develop mature reliable products, a reliability capability assessment and improvement process is needed.

As new companies and organizations emerge on the landscape, offering opportunities in new markets and new technological advances, their capabilities are unknown and their ability to deliver mature reliable high-quality products is uncertain. SIs, OEMs, PDs, and PMs need standards to help them determine how to assess organizations within their supply chains, especially organizations that are being considered as their new suppliers. These potential suppliers for SIs, OEMs, PDs, and PMs are assessed to ensure that they know what they are doing from a business sense as well as from a technological sense. Assessing the maturity of an organization is one way to determine if the organization knows what it is doing. Functional capability assessments (FCAs) are methods to determine the maturity of technological capabilities within organizations. One type of FCA is an organizational reliability capability assessment (ORCA). SIs, OEMs, PDs, and PMs set the requirements for system and product reliability, which may be derived from their customers' requirements, or based on market research, or flowed down directly from end-user customers to their supply chains. These companies benefit in using ORCA methodologies to separate the stellar organizations within their supply chains from the poorly developed organizations that are lacking maturity. IEEE 1624–2008 is a standard that assists these companies in developing an internal process that offers great value-added benefits in supplier selection and supply chain management related to organizational reliability capability.

THE BENEFITS OF IEEE 1624-2008

- IEEE 1624-2008 is a standard method used to assist designers in the selection of suppliers that includes assessment of the suppliers' capability to design and manufacture products meeting the customers' reliability requirements.
- IEEE 1624-2008 is also a standard method used to identify the shortcomings in reliability programs which can be rectified by subsequent improvement actions.

IEEE 1624–2008 explains the method used to assess the reliability capability of an organization, which is intended to be usable by all organizations that design, manufacture, or procure electrical and electronics components or products. The focus of this standard is on hardware products and system integrators, although the concepts described in this standard may be applied to both hardware and software products. The standard may be used for self-assessment by organizations or for assessing supplier capabilities for supplier–customer relationship development within a supply chain. The standard does not create an audit process but, rather, an assessment process that could be used to provide data and results as input into an audit process. The standard does not seek to create or propose creation of certifying bodies that audit whether a particular organization meets a defined level of reliability capability.

Electronic and mechanical parts and components, and contract manufacturing services purchased on the market as commodities, are selected based on information provided by suppliers and analyst reports concerning a particular industry segment. However, system integrators usually know very little about the reliability practices of their suppliers. Often, the organizations require that the suppliers prove reliability of the products by using inaccurate and outdated handbook-based reliability predictions as the first and principal way of measuring the expected reliability of products. It is only after they receive the parts or subassemblies that they can assess their reliability. This can be an expensive iterative process that has to be repeated for each new product. A solution to this problem is to identify the organizational core competencies that lead to highly reliable products. Companies should assess the suppliers within their supply chains that possess those core competencies that align with their vision for future system and product designs.

One of the reasons that reliability does not typically enter the decision-making process is the lack of an accepted methodology to measure the capability of an organization to develop and build reliable products. A supplier selection that takes into account the ability to meet reliability requirements can provide a valuable competitive advantage for the SIs, OEMs, PDs, and PMs. Evaluating the reliability activities of an electronics manufacturer can yield important information about the likelihood that the company will provide a reliable product.

Organizational reliability capability assessment can be performed at any stage in a product life cycle or even independent of any specific product. IEEE 15288-2008 [10] provides information on product life cycles. The assessment may be performed independent of any specific product. In the early stages of a product's life cycle, it is useful to carryout an internal organizational reliability capability assessment to determine if the organization has the capability to satisfy the product reliability requirements. Furthermore, it is useful to perform external organizational reliability capability assessments to determine if there is any technology that can satisfy the product reliability requirements. In the design stage, the organizational reliability capability assessments may focus more on specific products or manufacturing capability. In the production or support stage, reliability field data should be available so that an assessment can also be used as feedback on the effectiveness of various design and manufacturing changes.

ORGANIZATIONAL RELIABILITY CAPABILITY

Organizational reliability capability is the ability of an organization's reliability program, practices, or activities to assure that product reliability meets or exceeds its customers' requirements over an extended period of time. Furthermore, organizational reliability capability is the measure of the effectiveness of an organization's reliability in meeting or exceeding the customer's requirements for product reliability. Reliability capability is determined by the maturity levels of key practices which ensure that a supplier is capable of producing a reliable product, or which ensure that a reliability engineering service provider is capable of providing accurate reliability analysis and reliability testing results on a product. Reliability capability is a measure of the practices within an organization that contribute to the reliability of the final product and the effectiveness of these practices in meeting the reliability requirements of customers. The reliability capability of an organization may be a combination of the reliability capabilities of constituent reliability activities. Reliability capability assessment is the act of quantifying the effectiveness of these activities.

In the absence of systematic evaluation of these organizational traits, product testing and monitoring of field performance are often the primary means of assessing reliability over time. Improvement in product reliability based on these metrics is generally limited to a single product family and is difficult to institutionalize over the long term and across product lines. Periodic assessment of organizational reliability capability offers a method for identifying practices in need of improvement and implementing required improvements on a continual basis across multiple product lines or departments.

RELIABILITY CAPABILITY ASSESSMENT

Reliability capability assessment evaluates how well existing organizational traits meet defined objectives. The need for reliability capability assessment as a supply chain development tool has been recognized by many companies. IEEE 1624–2008 is the only standard that exists for defining the criteria for assessing organizational reliability capability. A reliability capability assessment can be performed by an external organization: for example, by another company seeking to establish a partner or supplier relationship. It can also be performed by independent consulting bodies hired for self-evaluation directly by an organization or acting as an agent for a prospective customer. An assessment may also be performed by an internal team as a normal business practice or as a response to a specific stimulus, such as customer complaints, excessive warranty costs, or a desire to use reliability to establish competitive advantage and market positioning.

Reliability capability assessment revolves around a set of key reliability practices that should be implemented in an organization to ensure delivery of reliable electronic products.

DESIGN CAPABILITY AND PERFORMABILITY

Design capability is the ability of the design to meet its performance requirements.

Performability is the means used to determine how reliably a design meets its performance requirements with adequate design margin. Design capability is demonstrated once during design qualification or design verification testing. Performability is demonstrated continuously over time, same as is done to determine reliability or dependability.

How Design Capability Affects Reliability Capability

A design capability may result in a probability of operation whose performance is less than optimal, due to the fundamental limits of the technology. When the design capability results in a less than optimal percentage, less than the operational requirements specified, the design capability has an adverse impact on the reliability. For example, if the design capability is stated in terms of the probability of a sensor detecting an object, and this design capability is stated as a requirement in the specification for the system (e.g., 70% probability of detection under benign ambient conditions), a design performance with a percentage of object detection less than the design capability requirement (e.g., 50% detection measured under test conditions) is a failure of the system that affects the system's reliability. That is based on the probability that the system will perform all the functions within acceptable levels to meet its requirements and to satisfy customer needs. The system may never fail due to hardware physical failure mechanisms over a certain period of time under specified operating environments and conditions, but if the performance of the design is not capable of meeting its requirements consistently, without a hardware failure, the reliability will be less than perfect. This is why design capability must be assessed as part of a reliability capability assessment.

IEEE 1332–1998 [9] identifies three reliability objectives to ensure that every reliability program activity adds value to the final system or product. The three reliability objectives defined in IEEE 1332–1998 are:

1. The supplier, working with the customer, should determine and understand the customer's requirements and product needs so that a comprehensive design specification can be generated.
2. The supplier should structure and follow a series of engineering activities that lead to a product that satisfies the customer's requirements and product needs with regard to reliability.
3. The supplier should include activities that assure the customer that reliability requirements and product needs have been satisfied.

These three objectives are achieved through implementation of eight key practices also described in IEEE 1624–2008 [10]. Figure 1 is a Venn diagram of the three reliability objectives composed of the eight key practices. Each practice is associated with several tasks, as discussed below.

1. *Reliability requirements (RR) practice* comprises the group of activities to evaluate customers' requirements, to generate reliability goals for products, to analyze and document requirements in specifications, and to plan reliability activities to meet the goals and requirements. The inputs for generating reliability requirements for products include customer inputs, reliability specifications, and lessons learned from experience on predecessor programs, systems, and products. Constraints such as budget, schedule, and technology maturity affect requirements and subsequent planning. The tasks of this practice may be implemented within the following life-cycle processes of IEEE 15288-2008 for acquisition, stakeholder requirements definition, requirements analysis, architectural design, project planning, project assessment and control, and measurement.

2. *Training and development (TD) practice* comprises a group of activities to enhance the technical, business, and specialized skills and knowledge of personnel so that they can perform their roles in manufacturing a reliable product effectively and efficiently. The aim is to ensure that employees understand the reliability plans and goals for products and to improve employee expertise in methods required to achieve those goals. This key practice also involves researching current practices, learning innovative techniques, and leveraging design or manufacturing technologies that can improve reliability. The tasks of this practice may be implemented within the following life-cycle process of IEEE 15288-2008 for human resource management.

3. *Reliability analysis (RA) practice* comprises a group of activities to conduct various types of reliability engineering analyses of system and product designs. These analysis tasks include the following as a minimum: (1) identifying stress conditions, (2) determining potential failure modes and failure mechanisms, and (3) providing reliability models, allocations, predictions, and assessments. The results of these analyses may include the quantification of risks for each component or subsystem.

4. *Reliability testing (RT) practice* comprises a group of activities to explore the design limits of a product and stress-screen products for design flaws, and to demonstrate the reliability of products. The tests may be conducted according to some industry standards or required customer specifications. The reliability testing procedures may be generic, or the tests may be custom-designed for each specific product. Detailed reliability test plans may include the test pass/fail criteria, stress conditions, sample size, and confidence levels.

5. *Supply-chain management (SM) practice* comprises a group of activities to identify sources of parts or processes that may be used to satisfy reliability requirements for a product and to manage suppliers (vendors and subcontractors) for long-term business association. Activities under this key practice include the tracking and approvals of product or part change notices, design reviews, and managing part obsolescence. These activities are essential for sustaining product reliability throughout the life cycle. They are also useful for making changes to product specifications, as well as for making design changes during a product's life cycle.

6. *Failure tracking (FT) practice* comprises a group of activities to collect manufacturing, yield, and field failure data. This failure tracking practice includes the method known as a failure reporting, analysis, and corrective action system. Failure data analysis is needed to analyze failures, identify the root causes of manufacturing defects and field failures, and generate failure analysis reports. The documented records for each report can include the date and lot code of the failed product, the failure point (quality testing, reliability testing, or field), the return date, the failure site, the failure mode and mechanism, and recommendations for avoiding the failure in existing and future products. For each product category, a Pareto chart of failure causes can be created and continually updated.

7. *Verification and validation (VV) practice* comprises the group of activities performed during the feedback process with the failure tracking practice. The verification and validation practice includes verification tasks through internal peer reviews or audits of reliability planning, testing, and analysis activities. These verification tasks ensure that planned reliability activities are implemented correctly during the design phase so that systems and products will be able to fulfill their specified reliability requirements. Validation tasks are performed to compare the reliability requirements, specifications, and reliability predictions that are completed during the design phase against the reliability data collected during design testing, production testing, and normal operation. Validation tasks determine the impact of reliability data on system and product specifications, and determine whether a specification change is needed or required. The field information and other empirical reliability data on systems and products are used to update reliability estimates, reliability test conditions, warranty cost estimates, and other logistics specifications, including spares provisioning, as defined in the reliability improvement practice.

8. *Reliability improvements (RI) practice* comprises a group of activities to implement corrective actions based on failure analysis and other results from the failure tracking and verification and validation practices. Tasks from the reliability improvements practice involve adhering to design change control processes for initiating design changes in systems and products. The practice will include tasks to update design processes as a result of changes in reliability requirements for systems and products, or changes in life-cycle application (operating and nonoperating) conditions of system and products.

Figure 1 is a Venn diagram of the three reliability objectives composed of the eight key practices.

An organization's ability to supply reliable products is quantified using a maturity-level metric. This multilevel metric allows a comparison of different organizations and also provides a baseline against which to measure an organization's improvement over time. The maturity approach to determining organizational abilities has roots in quality management. Crosby's quality management maturity grid [5] describes the typical behavior of an organization, which evolves through five phases (uncertainty, regression, awakening, enlightenment, and certainty) in its progression toward excellence. Since Crosby's grid was published, maturity models exist for a wide range of activities, including software development, supplier relationships, research and

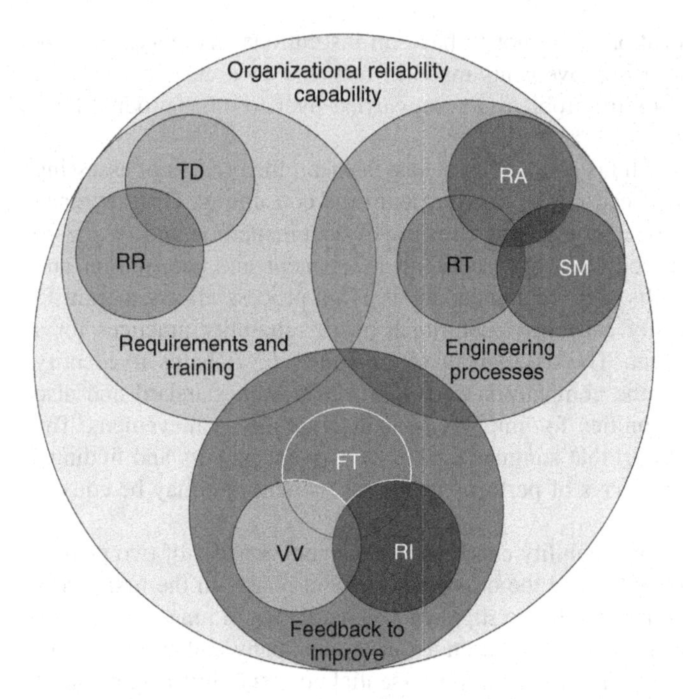

Figure 1 The eight key practices.

development effectiveness, collaboration, and product design. The most commonly used maturity models are the Capability Maturity Model (CMM) for software or the Capability Maturity Model Integration (CMMI) for systems. CMM and CMMI were developed by the Software Engineering Institute, which is part of Carnegie Mellon University. Both CMM and CMMI provide methods for assessing the capability of organizations to develop mature systems and products [6]. The reliability capability assessment model is analogous to the CMMI, focusing on systems and hardware products rather than software. The reliability capability maturity metric is defined as a measure of the practices within an organization that contribute to the reliability of the final product and the effectiveness of these practices in meeting customers' reliability requirements.

Although a set of key practices and associated reliability tasks are used in an assessment, reliability capability maturity is more than just performing a list of reliability-related tasks. Two organizations may have similar products and implement similar tasks. The more mature organization uses the tasks in an integrated fashion within the product life cycle. The mature organization implements tasks that provide value and reduce risk. A less mature organization may implement tasks only when required by a customer. By its nature, less mature organizations tend not to have an institutional memory, and any lessons learned from improvements made in response to a customer request or to address a reliability problem are not carried over to other product lines or future products.

Independent of who is conducting the assessment, the process of assessing reliability capability need not be onerous and time consuming. One approach that has been developed consists of a review of documentation and responses to a questionnaire, followed by an on-site assessment and preparation and presentation of results and recommendations. This process allows a team to determine a reliability capability maturity level of reliability practices for a facility or department. The assessment of capability level helps to identify those practices that the company is performing to a high standard and also indicates the opportunities for improvement in reliability achievement. The final product is a report that summarizes the assessment process and findings. There could be other ways of performing the assessment that may be equally effective.

The procedure for reliability capability assessment consists of two phases, the self-assessment phase and the on-site assessment phase. In the first phase, a questionnaire is submitted to the supplier which consists of nine subsections: one section on background information about the company and eight sections pertaining to each of the key practices essential to reliability achievement. The company is requested to identify the personnel who are best qualified to answer these questions and obtain their responses. The responses should be returned to the evaluators before the proposed on-site visit, with sufficient time to study the responses.

IEEE 1624 SCORING GUIDELINES

The IEEE 1624–2008 model contains five levels: the Solely Reactive, the Repeatable, the Defined, the Managed, and the Proactive. The Solely Reactive level features no formal organizational support for the capability in question. If anything is done, it is as a result of the efforts of individual personnel acting on their own initiative. There is no attempt to document anything relating to the reliability activity in question.

For a process to be classified as Repeatable, activities must be similar across product lines. Industry codes and standards are used as the basis for reliability plans, and the company can prove that its products conform to the relevant codes. However, actual comprehension of reliability concepts is minimal, and the process in question is implemented and followed in a very prescriptive fashion, with little allowance for the differences between products.

A level 3, or Defined, process is similar to that of level 2. However, there is much greater understanding among employees about exactly what they are doing and therefore more specific adaptation of the process to different products and situations. Processes at this level are usually better at establishing requirements for new designs than they are at providing feedback for current ones.

A level 4, or Managed, process allows the company to make changes to its existing products as a result of reliability data that it gathers. This process is documented and flexible and is used routinely during the development of a new product as well. In addition, the company involves its supply chain in the process as required.

Finally, the fifth, or Proactive, level identifies processes that are continuously improved as a result of experience and field data. All information gathered through the process is disseminated throughout the company. Lessons learned are incorporated into new and existing products.

SEI CMMI SCORING GUIDELINES

The SEI process also uses five levels of development for individual processes. The focus of the CMMI model is on the degree of visibility and control that management has over a given process. A higher level of control and visibility allows managers to track progress more accurately and gives earlier warning of deviations from the plan.

At level 1, a process is effectively a black box. There is no effective mechanism for managers to understand, track, or influence the process while it is under way. In the majority of cases, managers are unfamiliar with the technical nature of the process.

At level 2, the process can be viewed as a string of black boxes, with milestones placed between them. At this level, managers can see whether or not a particular subgoal has been reached on schedule, even if they cannot see how it is reached. In addition, the customer can review the product at defined milestones if desired.

At level 3, the tasks within the black boxes in level 2 are made visible. This is done by applying an organizational standard software process to the project. This standard process lists the tasks to be performed at a given stage in the process. All relevant employees understand what they are required to do, and management is able to plan for risks proactively.

At level 4, the tasks and processes identified in level 3 are controlled quantitatively. This allows a high degree of precision in planning, as the variability of the processes declines further. The customer is able to establish a clear, quantitative understanding of all relevant risks before a project begins.

At level 5, there is a formal system for process improvement. The process is constantly reworked as new ideas are developed or brought in from elsewhere in the organization. The customer is involved in the improvement efforts wherever possible.

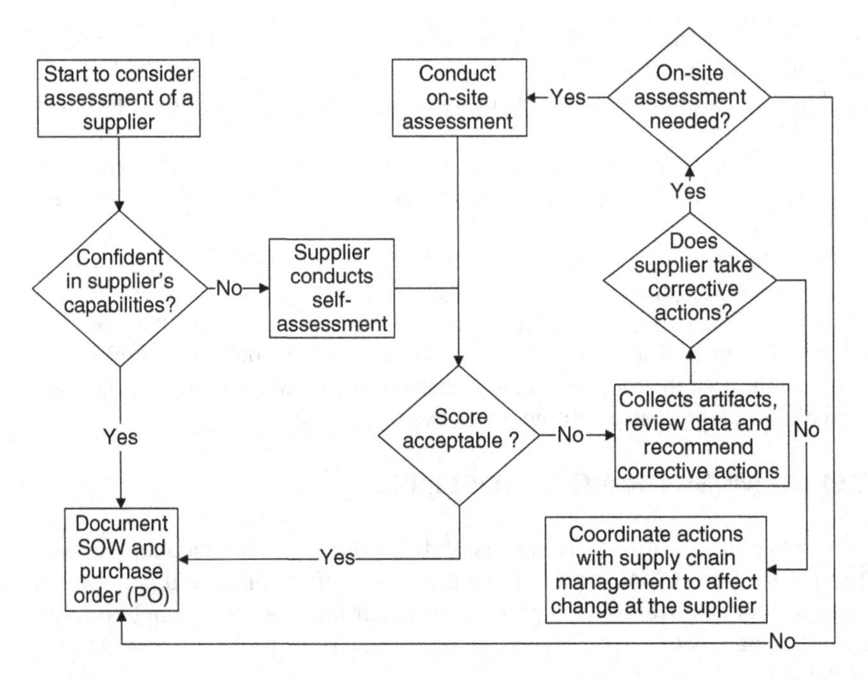

Figure 2 Example of a Process Flowchart.

ORGANIZATIONAL RELIABILITY CAPABILITY ASSESSMENT PROCESS

The ORCA process begins with the creation of a process document and a process flowchart. An example of a process flowchart is shown in Figure 2. The next step is to develop a process questionnaire with multiple-choice answers. The answers are scored from 0 to 5, with 5 being the best score. Each number in the questionnaire is defined by assessment criteria. The criteria can be selected from the IEEE 1624–2008 standard, or the SEI CMMI model, or a modification based on both scoring guidelines in IEEE and SEI.

Independent of who is conducting the assessment and for what purpose, the process of assessing reliability capability should not become onerous and time consuming. One approach consists of a review of documentation and responses to a questionnaire, followed by an on-site assessment and preparation and presentation of results and recommendations. This process allows a team to determine a reliability capability level of reliability practices for a facility or department. The assessment of capability level helps to identify those practices that the company is performing to a high standard and also indicates the opportunities for improvement in reliability achievement. The final product is a report that summarizes the assessment process and findings [11].

Another approach for reliability capability assessment consists of two phases: a self-assessment phase and an on-site assessment phase. In the first phase, a questionnaire is submitted to the supplier for their assessment of themselves. The questionnaire consists of nine subsections: one section on background information about the company and eight sections pertaining to each of the key practices essential to reliability achievement. Phase 1 ends when the supplier answers the questions and sends the questionnaire to the initiating organization. The responses to the self-assessment are reviewed and scored. This activity should be completed by the evaluators before the decision to conduct an on-site visit. Phase 2 involves forming an assessment team and traveling to the supplier to conduct the on-site assessment, using the same questionnaire as that provided to the supplier in phase 1. Phase 2 may be waived depending on the results of the phase 1 effort. There could be other ways of performing the assessment that may be equally effective [11].

Once the completed self-assessment form has been returned, the assessment leader will look through the scores. He or she will check to see if the scores are consistent with each other and that they are supported by references to company artifacts. If the assessment leader is not satisfied with the evidence offered in support of the scores, or if the scores are unexpectedly low, he or she will arrange an on-site assessment review. This will involve a meeting between the assessment leader and local supplier management to investigate the areas of uncertainty and determine what corrective actions need to be taken.

The evaluation team then assesses responses to the questionnaire and the supporting evidence, asking follow-up questions as necessary. At the conclusion of the meeting, the company is provided with an informal summary of the findings, including recommendations for corrective actions. The third and final phase involves documentation of the assessment. The company is provided with a draft report summarizing the evaluation team's observations and recommendations for reliability improvement. The company is typically given an opportunity to review the draft report and provide comments. A final report, which highlights the areas of strengths and weaknesses, with recommendations for improvements to approach best-in-class standards, is then issued to the company and to the organization that requested the assessment. The report also includes the maturity level of the company along with an explanation of the significance of that level.

Case Study 1 [7]

To assess the practicality of the reliability capability evaluation process, and as a part of an organization's reliability capability maturity development, a case study is presented. In the following paragraph we provide a brief profile of a company in terms of its reliability activities and describe the results of a reliability capability assessment.

This company is a leading manufacturer of electronic control products, providing thousands of products to customers in many countries. The warranty of the products usually ranges from one to two years, with a limited warranty of five years provided for some products. Most of their products are high-end products with specific reliability requirements, established based on past experience with similar products and customer feedback questionnaires. Reliability tasks are part of a quality plan, which is different for each business unit. A custom quality plan is generated for each product, keeping in view the requirements of the customer. Prior to implementation, the quality plan is reviewed by a cross-functional team, including people dealing with reliability. The company has reliability testing and failure analysis facilities, although some testing work is outsourced to leading test laboratories.

The company does not offer specific in-house training to its employees in broad areas of reliability. However, some employees have had outside training in specific topics, such as six sigma, the physics-of-failure approach, and highly accelerated life testing (HALT). The company conducts very limited failure modes and effects analysis for their product designs. They believe in designing systems and using parts that are tested to work beyond the expected usage cycles in the application environment. They "feel" that by adopting this approach, predicting reliability for their products becomes unnecessary. However, the company does have regular meetings with their service departments to inform them about potential component failures. Yearly meetings are also held to plan for reduction in field returns and component failure rates.

Most of the company's products are designed for a worst-case environment and a nominal 10-year useful life, and to have cumulative failures of less than a fraction of 1% over the life of the product. Most of the products are designed to internal specifications. Internal derating guidelines and thermal imaging are used in design. Materials used in product manufacture are also characterized for their heat resistance at elevated temperature usage. Any design changes made during a product development process are followed by requalification of the product. An internal product testing guideline has been developed to test a product design. The guideline incorporates tests, including HALT, temperature cycling, mechanical cycling, elevated temperature tests, maximum load testing, minimum load testing, and electrostatic discharge resistance tests. A standard series of tests is conducted for all products within a business unit. The company also conducts 100% end-of-line functional testing for their products.

A documented new product checklist is completed before any product goes into mass production. The company is proficient at understanding and monitoring life-cycle application conditions for its products. In some products, built-in software is used to assess usage. The company also conducts a simulation of the application and collects customer surveys to get the information. The purpose of these activities is to match application requirements with tests conducted. The company is currently also looking at methods for stress-health monitoring.

An approved vendor list is used for parts selection. This is accompanied by regular supplier audits conducted by the quality assurance group and statistical multiple-lot sample analysis of incoming parts and materials. The sample analysis includes mechanical and electrical testing. The selection of parts is generally made by the design group. The purchasing group is used only to keep track of the schedule and cost issues. Suppliers of critical parts are controlled directly by engineering. Otherwise, after initial selection, purchasing maintains control to ensure scheduled supplies. The company generally prefers to single-source parts, except for some commodity items that are multiple-sourced. The company very rarely uses parts outside their datasheet or supplier specifications. They use an internally maintained database to specify design ratings for supplied parts. All the parts used on existing products are approved for use on other products. Repeated "failures" of parts from a supplier will initiate action at the corporate level through the quality assurance department. The action can include exclusion of a supplier from future consideration. The company relies on its suppliers for testing of parts and for providing information about any product changes.

The company is currently in the process of developing a new system for assessing and updating the information about the impact of product change notices on their products. They believe in reengineering or redesigning their products and systems rather than relying on finding obsolete parts for older systems. The company uses a failure tracking system during and after manufacture. Manufacturing defects are tracked by corporate quality assurance, which may initiate a corrective action in some cases if defects rates are high. The postwarranty service and parts replacement provided by the company to their customers is noteworthy. Field failures are tracked even after the warranty period is over. Information of failures is obtained through a failure hotline, defective returns, and warranty returns. All failures tracked are included in a database providing information on the date of manufacture and date of return. However, shipping and sale dates are not tracked. All products that are returned from the field are analyzed. If a new failure mode is found, a new unit is subject to tests to reproduce the failure. The company uses the data from the field returns database to make improvements in its products by removing the failure causes or defective components. Field failures are tracked through successive generations of products to identify discrepancies. An improvement or deterioration initiates an investigation of the cause for the change. Some reliability tests have been redesigned based on field failures.

Case Study 2 [8]

An aircraft company has developed a model that is used to track the performance of each component, not only based on historical field history but also on what matters most to customers: the life-cycle cost. This very simple model enables the design and procurement teams to evaluate options and prioritize resources, and provides feedback very quickly to the design team during the design process as to which issues need to be addressed first.

The company describes the existing approach taken to improve the supply chain reliability performance and its weaknesses. The following paragraph introduces a new metric developed by the company to help prioritize the efforts of its engineering group and supply chain to focus on the issues that matter most to its customers and to help in designing reliability into components. This new procedure is part of the effort that the company has undertaken to achieve world-class performance.

It all starts with tracking the field performance of the components that are installed on the aircraft and providing the performance information back to the suppliers. The company has effective methods to gather and sort field unscheduled removal information to the component level. This not only provides the ability to track and monitor field reliability performance accurately but also permits the company's engineers and suppliers to quickly identify significant field issues. Each supplier receives field data directly on the performance and failures of its components. This unfiltered information permits suppliers to identify unusual patterns, to begin addressing emerging issues, and to truly partner with the company to identify and solve the underlying cause of current or potential field problems. The company uses an internally developed Web-based system called the supplier tracking and rating system (STARS) to provide performance feedback to suppliers relative to key supplier metrics. This feedback is provided to help each supplier identify and understand its performance relative to reliability, quality, cost, and schedule. The reliability rating is based on the total number of components installed on the company's systems and products which meet or exceed the reliability requirements established for each component. The rating and supporting details are provided to the supplier through the STARS electronic feedback process. These data are updated monthly and are based on the field removals as provided by the customers. The STARS process highlights those areas of reliability concern that require the supplier's participation in or initiation of the corrective action process.

ADVANTAGES OF HIGH RELIABILITY

The primary advantage [11] to an organization that is assessed at a high organizational reliability capability level is reliable products that result in lower warranty costs over the product warranty period. Lower warranty cost translates into higher profits and lower risk of product liabilities. Incremental cost reductions (ICRs) and one-time cost reductions (OCRs) are also advantages. These reductions may be passed on to the customer from lower warranty and manufacturing costs, due to improvements in the organizational reliability capability. As an organization's reliability processes get more and more mature, their products reach a high level of reliability maturity, so that life-cycle costs and total cost of ownership decrease over time. ICRs are the cost reductions that are realized incrementally over time as a product matures and optimal product learning curves are achieved. These may be passed on to customers as incentives to keep their business. An organization may be able to afford to offer ICRs due to their best practices associated with lower material and labor costs to produce the product with high reliability. OCRs are like ICRs except that there is a single instance where a cost reduction is realized and may be provided to the customer to ensure a strong business relationship. ICRs and OCRs may only be possible if the organizational reliability is capable enough to design and manufacture products that reach high levels of design maturity and high reliability of products is experienced early in the products' life cycles.

CONCLUSIONS

A reliability capability assessment process assists original equipment manufacturers, original design manufacturers, and system integrators in assessing prospective and existing suppliers. The assessment determines a supplier's ability to design and manufacture reliable and mature systems and products. Initially an organization employs this assessment process before suppliers deliver their products to the customer for use. The assessment process continues on a regular basis to help an organization and its supply chain identify gaps and weaknesses in their reliability program, which can be rectified by subsequent improvement actions. The assessment process establishes reliability management practices for use by designers, suppliers, customers, and independent authorities. The assessment method may be used to evaluate the reliability capability of any electronics-related organizations that perform activities influencing the reliability of a system or product. The assessment usually results in increased customer satisfaction, access to new competitive opportunities, and shortened system or product development cycle times.

In summary, a reliability capability assessment can be used to:

- Specify and plan reliability practices for system and product development
- Specify design ratings for supplier parts

- Evaluate reliability practices and tasks to determine the extent to which suppliers are capable of providing products that meet reliability requirements and reliability goals
- Improve reliability practices if the current reliability practices have been evaluated as deficient and improvement is desired or required
- Provide an appropriate and baseline checklist for quality assurance.

REFERENCES

[1] *IEEE Standard for Organizational Reliability Capability*, IEEE 1624–2008, IEEE, Piscataway, NJ, 2008.

[2] Gullo, L., Schenkelberg, F., Azarian, M., and Das, D., Assessment of organizational reliability capability, *IEEE Trans. Components Packag. Technol*., vol. 29, no. 2, June 2006.

[3] Pecht, M., and Ramakrishnan, A., Development and activities of the IEEE Reliability Standards Group, *J. Reliab. Eng. Assoc. Jpn*., vol. 22, no. 8, Nov. 2000, pp. 699–706.

[4] Tiku, S., and Pecht, M., Auditing the reliability capability of electronics manufacturers, *Adv. Electron. Packag*., vol. 1, 2003, pp. 947–953.

[5] Crosby, P. B., *Quality Is Still Free: Making Quality Certain in Uncertain Times*, McGraw-Hill, New York, 1996.

[6] Paulk, M. C., Weber, C. V., Garcia, S. M., Chrisis, M. B., and Bush, M., *Key Practices of the Capability Maturity Model, Version 1.1*, Tech. Rep. CMU/SEI-93-TR-025, ESC-TR-93-178, Software Engineering Institute, Carnegie Mellon University, Pittsburgh, PA, Feb. 1993.

[7] Tiku, S., Azarian, M., and Pecht, M., CALCE Electronic Products and Systems Center, University of Maryland, College Park, Maryland, USA, Using a reliability capability maturity model to benchmark electronics companies, *Int. J. Quality Reliab. Manage*., vol. 24 no. 5, 2007, pp. 547–563.

[8] Hamada, M., and Jarrell, G., World-class supply chain reliability in general aviation, in *Annual Reliability and Maintainability Symposium 2008 Proceedings*.

[9] *IEEE Standard Reliability Program for the Development and Production of Electronics Systems and Equipment*, IEEE 1332–1998, IEEE, Piscataway, NJ, June 1998.

[10] *IEEE Systems and Software Engineering: System Life Cycle Processes*, IEEE 15288-2008, IEEE, Piscataway, NJ, Jan. 2008.

[11] Gullo, L., Advantages of IEEE P1624 for assessing organizational reliability capability, in *Annual Reliability and Maintainability Symposium 2009 Proceedings*.

Index

Design for Reliability, First Edition. Edited by Dev Raheja, Louis J. Gullo.

Printed in the United States
By Bookmasters